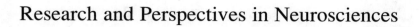

Research and Perspectives in Neurosciences

For further volumes:
http://www.springer.com/series/2357

Fred H. Gage • Yves Christen

Editors

Programmed Cells from Basic Neuroscience to Therapy

 Springer

Editors
Fred H. Gage
Laboratory of Genetics
The Salk Institute f. Biological Studies
La Jolla, California
USA

Yves Christen
Fondation Ipsen
Boulogne Billancourt
France

ISSN 0945-6082
ISBN 978-3-642-36647-5   ISBN 978-3-642-36648-2 (eBook)
DOI 10.1007/978-3-642-36648-2
Springer Heidelberg New York Dordrecht London

Library of Congress Control Number: 2013938743

Printed on acid-free paper

Springer is part of Springer Science+Business Media (www.springer.com)

Programmed Cells from Basic Neuroscience to Therapy

Studies of human brain and neuronal function in phenotypically normal as well as neurological and psychiatric patients have been performed using noninvasive imaging methods. However, the spatial and temporal limitations do not permit single cell/neuron resolution. In addition, genomic and molecular studies of neurological and psychiatric patients are conducted on postmortem tissues often representing the end-stage of life and disease or from peripheral tissues and biopsies, and blood. The recent advances in programming somatic cells (PSC), including induced pluripotent stem cells (iPS) and induced neuronal phenotypes (iN), have changed the experimental landscape and opened new possibilities. These advances have provided an important tool for the study of human neuronal function as well as neurodegenerative and neurodevelopmental diseases in live human neurons in a controlled environment. Researchers are just beginning to take advantage of the many implications of studying developing neurons from living humans in vitro. For example, reprogramming cells from patients with neurological diseases allows the study of molecular pathways particular to specific subtypes of neurons, such as dopaminergic neurons in Parkinson's disease, motor neurons for amyolateral sclerosis, or myelin for multiple sclerosis. In addition, because PSC technology allows for the study of human neurons during development, disease-specific pathways can be investigated prior to and during disease onset. Detecting disease-specific molecular signatures in live human brain cells opens possibilities for early intervention therapies and new diagnostic tools. Importantly, it is now feasible to obtain gene expression profiles from neurons that capture the genetic uniqueness of each patient. Importantly, once the neurological neural phenotype is detected in vitro, the so-called disease-in-a-dish approach allows for the screening of drugs that can ameliorate the disease-specific phenotype. New therapeutic drugs could either act on generalized pathways in all patients or be patient-specific and used in a personalized medicine approach. However, there are a number of pressing issues that need to be addressed and resolved before PSC technology can be extensively used for clinically relevant modeling of neurological diseases. Among these issues are the variability in PSC generation methods, variability between individuals, epigenetic/genetic instability, and the ability to obtain disease-relevant subtypes of neurons. Current protocols for differentiating PSC into specific subtypes of neurons are under development, but more and better protocols are needed. Understanding the molecular pathways involved in human neural differentiation will facilitate the development of methods and tools to enrich and monitor the generation of specific subtypes of neurons that would be more relevant in modeling different neurological diseases. The meeting of Fondation IPSEN on "Programmed cells: from basic neuroscience to therapy" held in Paris, April 2, 2012, is well captured in this volume and reflects the cautious optimism exhibited by the participants of the meeting.

Fred H. Gage
Yves Christen

Acknowledgments

The editors wish to express their gratitude to Mrs. Mary Lynn Gage for her editorial assistance and Mrs. Jacqueline Mervaillie and Sonia Le Cornec for the organization of the meeting.

Contents

Contributors

Juan Carlos Izpisua Belmonte Gene Expression Laboratory, The Salk Institute for Biological Studies, La Jolla, CA, USA

Oliver Brüstle Institute of Reconstructive Neurobiology, Life & Brain Center, University of Bonn and Hertie Foundation, Bonn, Germany

Elena Cattaneo Centre for Stem Cell Research, Università degli Studi di Milano, Milan, Italy

Gist F. Croft Project A.L.S./Jenifer Estess Laboratory for Stem Cell Research, Columbia Stem Cell Initiative, Center for Motor Neuron Biology and Disease, Departments of Rehabilitation and Regenerative Medicine, Pathology and Cell Biology, Neurology and Neuroscience, Columbia University Medical Center, New York, NY, USA

Fred H. Gage Laboratory of Genetics, The Salk Institute for Biological Studies, La Jolla, CA, USA

Lawrence S.B. Goldstein Department of Cellular and Molecular Medicine and Department of Neurosciences, Sanford Consortium for Regenerative Medicine and UCSD School of Medicine, La Jolla, CA, USA

John B. Gurdon Wellcome Trust/Cancer Research UK Gurdon Institute, University of Cambridge, Cambridge, UK

Department of Zoology, University of Cambridge, Cambridge, UK

Christopher E. Henderson Project A.L.S./Jenifer Estess Laboratory for Stem Cell Research, Columbia Stem Cell Initiative, Center for Motor Neuron Biology and Disease, Departments of Rehabilitation and Regenerative Medicine, Pathology and Cell Biology, Neurology and Neuroscience, Columbia University Medical Center, New York, NY, USA

Rudolf Jaenisch Whitehead Institute, Massachusetts Institute of Technology, Cambridge, MA, USA

Philipp Koch Institute of Reconstructive Neurobiology, Life & Brain Center, University of Bonn and Hertie Foundation, Bonn, Germany

Maria C. N. Marchetto Laboratory of Genetics, The Salk Institute for Biological Studies, La Jolla, CA, USA

Cécile Martinat INSERM/UEVE 861, I-Stem, AFM, Evry, France

Jerome Mertens Institute of Reconstructive Neurobiology, Life & Brain Center, University of Bonn and Hertie Foundation, Bonn, Germany

Alysson Renato Muotri School of Medicine, Department of Pediatrics/Rady Children's Hospital San Diego, Department of Cellular & Molecular Medicine, Stem Cell Program, University of California San Diego, La Jolla, CA, USA

Emmanuel Nivet Gene Expression Laboratory, Salk Institute for Biological Studies, La Jolla, CA, USA

Derek H. Oakley Project A.L.S./Jenifer Estess Laboratory for Stem Cell Research, Columbia Stem Cell Initiative, Center for Motor Neuron Biology and Disease, Departments of Rehabilitation and Regenerative Medicine, Pathology and Cell Biology, Neurology and Neuroscience, Columbia University Medical Center, New York, NY, USA

Marc Peschanski INSERM/UEVE 861, I-Stem, AFM, Evry, France

Ignacio Sancho-Martinez Gene Expression Laboratory, Salk Institute for Biological Studies, La Jolla, CA, USA

Stan Wang Wellcome Trust/Cancer Research UK Gurdon Institute and Department of Surgery, School of Clinical Medicine, University of Cambridge, Cambridge, United Kingdom

Marius Wernig Institute for Stem Cell Biology and Regenerative Medicine, Department of Pathology, Stanford University School of Medicine, Stanford, CA, USA

Hynek Wichterle Project A.L.S./Jenifer Estess Laboratory for Stem Cell Research, Columbia Stem Cell Initiative, Center for Motor Neuron Biology and Disease, Departments of Rehabilitation and Regenerative Medicine, Pathology and Cell Biology, Neurology and Neuroscience, Columbia University Medical Center, New York, NY, USA

Chiara Zuccato Centre for Stem Cell Research, Università degli Studi di Milano, Milan, Italy

iPS Cell Technology and Disease Research: Issues To Be Resolved

Rudolf Jaenisch

Abstract The ability to reprogram somatic cells into induced pluripotent stem (iPS) cells has opened the possibility of studying human diseases in the Petri dish and may eventually allow the treatment of major diseases by customized cell therapy. This review summarizes open questions in the reprogramming field that need to be resolved to make iPS technology a robust and general approach for the study of human disease. The experimental challenges that need to be worked out include establishing efficient gene targeting methods that allow the generation of genetically defined, disease-specific and control cells, the development of robust differentiation protocols and the production of non genetically altered iPS cells.

Modifying the original protocol established by Takashi and Yamanaka, patient-specific iPS cells have been generated from a variety of donor cell types (Jaenisch and Young 2008; Yamanaka 2007). The methods to induce somatic cells to pluripotent iPS cells include the transduction of the reprogramming factors by retro- or lentiviral vectors, non-integrating or excisable vectors. The reprogramming process is characterized by widespread epigenetic changes (Kim et al. 2010; Mikkelsen et al. 2008) that generate iPS cells that are functionally and molecularly similar to embryonic stem (ES) cells.

Given these data, a stochastic model has emerged to explain how forced expression of the transcription factors initiates the process that eventually leads to the pluripotent state (Hanna et al. 2009, 2010b; Yamanaka 2009). The stochastic model supposes that the reprogramming factors in the somatic cells initiate a sequence of epigenetic events that eventually lead to the small and unpredictable fraction of iPS cells. A number of experimental parameters have been shown to affect the quality

R. Jaenisch (✉)
Whitehead Institute, Massachusetts Institute of Technology, 9 Cambridge Center, Cambridge, MA 02142, USA
e-mail: jaenisch@wi.mit.edu

F.H. Gage and Y. Christen (eds.), *Programmed Cells from Basic Neuroscience to Therapy*, Research and Perspectives in Neurosciences 20,
DOI 10.1007/978-3-642-36648-2_1, © Springer-Verlag Berlin Heidelberg 2013

of iPS cells, including the choice of vectors, the stochiometry of the reprogramming factors (Carey et al. 2011) and particular growth conditions (Hanna et al. 2010a).

A major concern for the study of human diseases in the cell culture dish is that it may be difficult or not practical to model in vitro diseases with a long-latency such as Alzheimer's or Parkinson's disease. In such cases, the dynamics of disease progression in the patient are likely to be vastly different from any phenotype developing in vitro in cells differentiated from patient-specific human iPS (hiPS) cells. It may be possible to accelerate the development of pathological phenotypes in cell culture by exposing the cultures to environmental insults that are suspected or known to contribute to the disease (Saha and Jaenisch 2009). Such treatments may consist of insults such as oxidative stress, pesticide or heavy metal exposure. Also, any non-autonomous diseases that involve the interaction of several cell types may be difficult to model in vitro with a single, purified, lineage-committed cell type unless tissues culture systems can be developed that allow the study of the interaction of different cell types in the Petri dish. Finally, because the reprogramming process is expected to remove any epigenetic alterations associated with disease phenotypes (Jaenisch and Bird 2003), any environmental effects that may contribute to a disease would be difficult to study in functional cells derived from patient-derived iPS cells. This issue is particularly relevant to sporadic and multifactorial disorders caused by a combination of genetic and environmental factors. Thus, iPS cells from patients with sporadic diseases, which may be caused predominantly by epigenetic alterations, may be of little value for mechanistic studies unless the epigenetic alterations also associate with unidentified genetic alterations.

In this chapter, I will address two technical challenges: (1) the creation of reprogramming factor-free hiPS cells to minimize or eliminate genetic alterations in the derived iPS cell lines; and (2) gene targeting strategies to generate (i) markers for differentiation and gene corrections and (ii) strategies to produce isogenic pairs of disease-specific and control cells.

Strategies for Deriving Reprogramming Factor-Free hiPS Cells

One concern is that constitutive expression of integrated copies of reprogramming factors could influence the differentiation potential and growth properties of iPS cells in culture or after transplantation. For example, any expression of c-Myc will increase the risk of tumor development, and a comparison of iPS cells before or after excision of the reprogramming vectors revealed global gene expression changes, indicating that even low residual vector expression could profoundly affect the biological properties of iPS cells (Soldner et al. 2009). Thus, it is of paramount importance to generate iPS cells that carry no genetic alterations. Four different strategies have been used to generate factor-free iPS cells.

1. *Non integrating or excisable vectors:* The use of a DOX inducible vector that also carries LoxP sites is an approach to efficiently induce iPS cells and to excise

the vectors by transient Cre expression (Soldner et al. 2009). However, the loxP sites will remain after excision of the vectors. Another approach is the use of the piggyBac transposition system, which allows removal of exogenous reprogramming factors from genomic integration sites in iPS cells (Woltjen et al. 2009). Non-integrating episomal vectors have also been used to reprogram cells (Yu et al. 2009). Though all of these approaches achieve removal of the vectors, the use of DNA transfection could lead to random integrations of vector sub-fragments that may be difficult to detect. Finally, Sendai viral vectors have been used to induce reprogramming (Seki et al. 2010). While this approach is efficient, it is important to ascertain that the virus does not persist in the iPS cells.

2. *Protein or RNA transfection:* The introduction of the reprogramming factors by non-DNA based methods has been successful. While protein transfection of the factors has been reported, the efficiency of the approach to induce iPS cells was so low that this approach had no practical value (Kim et al. 2009). An interesting method to obtain genetically unmodified iPS cells is the use of mRNA encoding the reprogramming factors (Warren et al. 2010). While the approach involves repeated RNA transfection, it has now been used to generate iPS cells from a verity of donor cells.

3. *Small molecules:* A variety of small molecules have been shown to replace reprogramming factors (Huangfu et al. 2008; Lyssiotis et al. 2009), but so far reprogramming has not been achieved with only small molecules.

Genetic Modification of hES Cells and hiPS Cells

Gene targeting by homologous recombination has proven to be inefficient in hES and hiPS cells, which has hampered the development of tools that are essential to realize the full potential of ES and iPS cells for disease research. I will briefly summarize the development and use of new technologies that are helping to overcome these limitations.

Recently, site-specific zinc-finger nucleases (ZFNs) have been shown to facilitate homologous recombination (Lombardo et al. 2007). A ZFN is generated by fusing the FokI nuclease domain to a DNA recognition domain composed of engineered zinc-finger motifs that specify the genomic DNA binding site for the chimeric protein. Upon binding of two such fusion proteins at adjacent genomic sites, the nuclease domains dimerize, become active and cut the genomic DNA. When a donor DNA that is homologous to the target on both sides of the double-strand break is provided, the genomic site can be repaired by homology-directed repair, allowing the incorporation of exogenous sequences placed between the homologous regions. ZNF gene editing has been used to target endogenous genes in hES cells and hiPS cells (Hockemeyer et al. 2009), allowing efficient insertion of markers such as GFP into endogenous genes. While this approach has been successful to target expressed and silent genes, the design of ZNFs is complex. More recently, Transcription Activator Like Effector Nucleases (TALENs) have been used as an alternative gene editing approach (Hockemeyer et al. 2011). While as efficient as ZFNs in targeting genes, the advantage of the TALEN-mediated gene editing is that these nucleases are easy to design and can be generated within a few days in the laboratory. The following applications of these targeting approaches for disease research are envisaged.

Markers for Differentiation

One of the major unresolved issues of the stem cell field is the inefficiency and poor reproducibility of differentiating pluripotent cells into functional differentiated cells that could identify an in vitro disease-relevant phenotype or could be used as donors for cell transplantation. It would, therefore, be desirable to develop indicator cell lines that carry a GFP gene inserted into key transcription factors the activation of which could be used as convenient markers to develop robust differentiation protocols. To assure robust marker expression under the control of the endogenous gene without interfering with its function, we have, using TALEN-mediated gene editing, inserted the GFP sequences into several endogenous genes $3'$ of the stop coding under the control of 2A sequences (Hockemeyer et al. 2011). This strategy resulted in marker lines that exhibited robust and faithful marker expression in the OCT4 gene, the PPP1R12C gene (the common AAVS1 integration site) and the PITX3 gene. Thus, this strategy may be a general approach to develop marker cell lines that could be used for screening chemical libraries in efforts to identify small molecule compounds that drive ES cells into particular differentiation pathways.

Generation of Isogenic Pairs of Disease-Specific and Control Cells

One of the most immediate and exciting applications of iPS technology is establishing disease models in the Petri dish. However, a potentially serious complication of using iPS or ES cells for disease research is the variability of differentiating cells to a desired and disease-relevant phenotype in vitro. Thus, any phenotype discovered in disease-specific cells as compared to cells from a normal donor could be due to system-imminent variability rather than a disease-relevant effect. The basis for this variability is manifold and includes (1) differences in genetic background; (2) the process of cell derivation (Carey et al. 2011; Lengner et al. 2010) and (3) in the case of hiPS cells, variegation effects and residual transgene expression of the viral vectors used to induce reprogramming (Soldner et al. 2009) and genetic alterations introduced during the reprogramming process (Gore et al. 2011; Hussein et al. 2011).

While in vitro models of early-age-onset or metabolic diseases are more likely to display clear differences when compared to healthy donor controls, late-age-onset disorders such as Parkinson's and Alzheimer's disease, with long latency and slow progression of cellular and pathological changes in vivo, are expected to show only subtle if any informative phenotypes in the Petri dish. Thus, it is of great importance to distinguish subtle but disease-relevant phenotypical changes from unpredictable experimental genetic background-related or system-imminent variability due to the lack of genetically matched controls. Commonly used control cells are derived from healthy donors but pose major problems because individual hES and hiPS cell lines display highly variable biological characteristics, such as the propensity to differentiate into specific functional cells (Bock et al. 2011; Huangfu et al. 2008). Therefore, for the "disease in a dish" approach to be successful, it is essential to set up experimental systems in which the disease-causing genetic lesion of interest is the sole modified variable. We have used the ZNF technology to generate isogenic disease and control cell lines from hES and hiPS cells that differ exclusively at well-validated susceptibility variants for Parkinson's disease by genetically modifying single base pairs in the α-synuclein gene.

Mutations in α-synuclein such as A53T, E46K, A30P are known to lead to early-onset Parkinson's disease. To develop a genetically defined, human in vitro model of Parkinson's disease, we generated a panel of control and disease-related cell lines by either deriving hiPS cells from a patient carrying the A53T (G209) α-synuclein mutation followed by the correction of this mutation or, alternatively, by generating either the A53T (G209A) or E46K (G188A) mutation in the genome of wild-type hES cells (Soldner et al. 2011). Any alteration seen in neurons derived from the mutant lines that was different from that seen in neurons derived from the genetically matched control cell lines would indicate(as meant?) that the phenotype is disease-relevant rather than due to uncontrollable differences in genetic background or other system-imminent variability.

Outlook

It is likely that the generation of patient-specific iPS cells will have a significant impact on the study of human diseases and on regenerative medicine. However, as outlined in this brief review, a number of technical issues need to be resolved before the technology can be used in a clinical setting (Saha and Jaenisch 2009). These include the establishment of efficient reprogramming strategies that do not result in genetically modified cells and of genetically defined disease and control cells. One of the key challenges for translating these new technologies to the clinic is devising robust protocols for differentiating ES or iPS cells into self-renewing adult stem cells and lineage-committed cells. Armed with such protocols, researchers can begin to define experimental conditions that allow the development and detection of relevant in vitro phenotypes for a given human disease, putting "personalized" regenerative medicine on the horizon.

References

Bock C, Kiskinis E, Verstappen G, Gu H, Boulting G, Smith ZD, Ziller M, Croft GF, Amoroso MW, Oakley DH, Kiskinis E, Verstappen G, Gu H, Boulting G, Smith ZD, Ziller M, Croft GF, Amoroso MW, Oakley DH, Gnirke A, Eggan K, Meissner A (2011) Reference maps of human ES and iPS cell variation enable high-throughput characterization of pluripotent cell lines. Cell 144:439–452

Carey BW, Markoulaki S, Hanna JH, Faddah DA, Buganim Y, Kim J, Ganz K, Steine EJ, Cassady JP, Creyghton MP, Welstead GG, Gao Q, Jaenisch R (2011) Reprogramming factor stoichiometry influences the epigenetic state and biological properties of induced pluripotent stem cells. Cell Stem Cell 9:588–598

Gore A, Li Z, Fung HL, Young JE, Agarwal S, Antosiewicz-Bourget J, Canto I, Giorgetti A, Israel MA, Kiskinis E, Lee JH, Loh YH, Manos PD, Montserrat N, Panopoulos AD, Ruiz S, Wilbert ML, Yu J, Kirkness EF, Izpisua Belmonte JC, Rossi DJ, Thomson JA, Eggan K, Daley GQ, Goldstein LS, Zhang K (2011) Somatic coding mutations in human induced pluripotent stem cells. Nature 471:63–67

Hanna J, Saha K, Pando B, van Zon J, Lengner CJ, Creyghton MP, van Oudenaarden A, Jaenisch R (2009) Direct cell reprogramming is a stochastic process amenable to acceleration. Nature 462:595–601

Hanna J, Cheng AW, Saha K, Kim J, Lengner CJ, Soldner F, Cassady JP, Muffat J, Carey BW, Jaenisch R (2010a) Human embryonic stem cells with biological and epigenetic characteristics similar to those of mouse ESCs. Proc Natl Acad Sci USA 107:9222–9227

Hanna JH, Saha K, Jaenisch R (2010b) Pluripotency and cellular reprogramming: facts, hypotheses, unresolved issues. Cell 143:508–525

Hockemeyer D, Wang H, Kiani S, Lai CS, Gao Q, Cassady JP, Cost GJ, Zhang L, Santiago Y, Miller JC et al (2011) Genetic engineering of human pluripotent cells using TALE nucleases. Nat Biotechnol 29:731–734

Hockemeyer D, Soldner F, Beard C, Gao Q, Mitalipova M, DeKelver RC, Katibah GE, Amora R, Boydston EA, Zeitler B, Meng X, Miller JC, Zhang L, Rebar EJ, Gregory PD, Urnov FD, Jaenisch R (2009) Efficient targeting of expressed and silent genes in human ESCs and iPSCs using zinc-finger nucleases. Nat Biotechnol 27:851–857

Huangfu D, Maehr R, Guo W, Eijkelenboom A, Snitow M, Chen AE, Melton DA (2008) Induction of pluripotent stem cells by defined factors is greatly improved by small-molecule compounds. Nat Biotechnol 26:795–797

Hussein SM, Batada NN, Vuoristo S, Ching RW, Autio R, Narva E, Ng S, Sourour M, Hamalainen R, Olsson C, Lundin K, Mikkola M, Trokovic R, Peitz M, Brüstle O, Bazett-Jones DP, Alitalo K, Lahesmaa R, Nagy A, Otonkoski T (2011) Copy number variation and selection during reprogramming to pluripotency. Nature 471:58–62

Jaenisch R, Bird A (2003) Epigenetic regulation of gene expression: how the genome integrates intrinsic and environmental signals. Nat Genet 33(Suppl):245–254

Jaenisch R, Young R (2008) Stem cells, the molecular circuitry of pluripotency and nuclear reprogramming. Cell 132:567–582

Kim D, Kim CH, Moon JI, Chung YG, Chang MY, Han BS, Ko S, Yang E, Cha KY, Lanza R, Kim KS (2009) Generation of human induced pluripotent stem cells by direct delivery of reprogramming proteins. Cell Stem Cell 4:472–476

Kim K, Doi A, Wen B, Ng K, Zhao R, Cahan P, Kim J, Aryee MJ, Ji H, Ehrlich LI, Yabuuchi A, Takeuchi A, Cunniff KC, Hongguang H, McKinney-Freeman S, Naveiras O, Yoon TJ, Irizarry RA, Jung N, Seita J, Hanna J, Murakami P, Jaenisch R, Weissleder R, Orkin SH, Weissman IL, Feinberg AP, Daley GQ (2010) Epigenetic memory in induced pluripotent stem cells. Nature 467:285–290

Lengner CJ, Gimelbrant AA, Erwin JA, Cheng AW, Guenther MG, Welstead GG, Alagappan R, Frampton GM, Xu P, Muffat J, Santagata S, Powers D, Barrett CB, Young RA, Lee JT,

Jaenisch R, Mitalipova M (2010) Derivation of pre-X inactivation human embryonic stem cells under physiological oxygen concentrations. Cell 141:872–883

Lombardo A, Genovese P, Beausejour CM, Colleoni S, Lee YL, Kim KA, Ando D, Urnov FD, Galli C, Gregory PD, Holmes MC, Naldini L (2007) Gene editing in human stem cells using zinc finger nucleases and integrase-defective lentiviral vector delivery. Nat Biotechnol 25:1298–1306

Lyssiotis CA, Foreman RK, Staerk J, Garcia M, Mathur D, Markoulaki S, Hanna J, Lairson LL, Charette BD, Bouchez LC, Bollong M, Kunick C, Brinker A, Cho CY, Schultz PG, Jaenisch R (2009) Reprogramming of murine fibroblasts to induced pluripotent stem cells with chemical complementation of Klf4. Proc Natl Acad Sci USA 106:8912–8917

Mikkelsen TS, Hanna J, Zhang X, Ku M, Wernig M, Schorderet P, Bernstein BE, Jaenisch R, Lander ES, Meissner A (2008) Dissecting direct reprogramming through integrative genomic analysis. Nature 454:49–55

Saha K, Jaenisch R (2009) Technical challenges in using human induced pluripotent stem cells to model disease. Cell Stem Cell 5:584–595

Seki T, Yuasa S, Oda M, Egashira T, Yae K, Kusumoto D, Nakata H, Tohyama S, Hashimoto H, Kodaira M, Okada Y, Seimiya H, Fusaki N, Hasegawa M, Fukuda K (2010) Generation of induced pluripotent stem cells from human terminally differentiated circulating T cells. Cell Stem Cell 7:11–14

Soldner F, Hockemeyer D, Beard C, Gao Q, Bell G, Cook E, Hargus G, Cooper ABO, Mitalipova M, Isacson O, Jaenisch R (2009) Parkinson's disease patient-derived induced pluripotent stem cells free of viral reprogramming factors. Cell 136:964–977

Soldner F, Laganiere J, Cheng AW, Hockemeyer D, Gao Q, Alagappan R, Khurana V, Golbe LI, Myers RH, Lindquist S, Zhang L, Guschin D, Fong LK, Vu BJ, Meng X, Urnov FD, Rebar EJ, Gregory PD, Zhang HS, Jaenisch R (2011) Generation of isogenic pluripotent stem cells differing exclusively at two early onset Parkinson point mutations. Cell 146:318–331

Warren L, Manos PD, Ahfeldt T, Loh YH, Li H, Lau F, Ebina W, Mandal PK, Smith ZD, Meissner A, Daley GQ, Brack AS, Collins JJ, Cowan C, Schlaeger TM, Rossi DJ (2010) Highly efficient reprogramming to pluripotency and directed differentiation of human cells with synthetic modified mRNA. Cell Stem Cell 7:618–630

Woltjen K, Michael IP, Mohseni P, Desai R, Mileikovsky M, Hamalainen R, Cowling R, Wang W, Liu P, Gertsenstein M, Kaji K, Sung HK, Nagy A (2009) piggyBac transposition reprograms fibroblasts to induced pluripotent stem cells. Nature 458:766–770

Yamanaka S (2007) Strategies and new developments in the generation of patient-specific pluripotent stem cells. Cell Stem Cell 1:39–49

Yamanaka S (2009) Elite and stochastic models for induced pluripotent stem cell generation. Nature 460:49–52

Yu J, Hu K, Smuga-Otto K, Tian S, Stewart R, Slukvin II, Thomson JA (2009) Human induced pluripotent stem cells free of vector and transgene sequences. Science 324:797–801

Therapeutic Somatic Cell Reprogramming by Nuclear Transfer

Stan Wang and John B. Gurdon

Abstract In the course of normal development, cells rarely are able to revert from a differentiated state back to an embryonic state. However, techniques exist that allow this reversal to take place. In an experiment performed over 50 years ago, single cell nuclear transfer from somatic cells to enucleated eggs was able to yield successful development of cloned *Xenopus laevis* (Gurdon et al., Nature 182:64–65, 1958). Through somatic cell nuclear transfer (NT), several cell divisions occur before the onset of new gene transcription; moreover, new cell types and even organisms can be derived (Campbell et al., Nature 380:64–66, 1996). More recently, terminally differentiated cells could be induced to reprogram to a pluripotent, embryonic stem (ES) cell-like state via overexpression of a particular subset of transcription factors (TF) (Takahashi and Yamanaka, Cell 126:663–676, 2006). These induced pluripotent stem (iPS) cells can then be re-differentiated into various tissue types, including both somatic and germ cells. A possible advantage that somatic cell NT harbors over iPS is that factors present in the egg have been shown to directly remove silencing of genes via chromatin decondensation, removal of histone modifications, and activation of gene transcription prior to cell division. Therefore, an improved understanding of how the egg facilitates nuclear reprogramming by natural means may identify components that can be used for more efficient reprogramming by this and other means.

S. Wang
Wellcome Trust/Cancer Research UK Gurdon Institute, University of Cambridge, Cambridge, UK

Department of Surgery, School of Clinical Medicine, University of Cambridge, Cambridge, UK

J.B. Gurdon (⊠)
Wellcome Trust/Cancer Research UK Gurdon Institute, University of Cambridge, Cambridge, UK

Department of Zoology, University of Cambridge, The Henry Wellcome Building of Cancer and Developmental Biology, Tennis Court Road, Cambridge CB2 1QN, UK
e-mail: j.gurdon@gurdon.cam.ac.uk

F.H. Gage and Y. Christen (eds.), *Programmed Cells from Basic Neuroscience to Therapy*, Research and Perspectives in Neurosciences 20,
DOI 10.1007/978-3-642-36648-2_2, © Springer-Verlag Berlin Heidelberg 2013

Therapeutic Limitations of Reprogramming Techniques

Although iPS cells hold great promise for the generation of patient-specific pluripotent stem cells, several challenges currently exist that limit their direct application in human therapy. Though increasing with recent techniques, the efficiency of generating iPS cells remains low (Wang et al. 2011; Chen et al. 2011). In addition, iPS cells can harbor increased tumorigenic potential due to the use of genome-incorporating viruses in the original reprogramming process, along with having an increased presence of oncogenes (Zhang et al. 2012). This issue has since been remedied by implementation of integration-free methods, such as episomal plasmid vectors (Okita et al. 2011). Furthermore, iPS cells have been demonstrated to have issues regarding increased copy number variation, somatic mutations, and aberrant epigenetics (Hussein et al. 2011; Gore et al. 2011; Lister et al. 2011). However, some recent data indicate that these could be the result of abnormalities already present in the original cell lines (Young et al. 2012).

Nuclear reprogramming via NT provides a few distinct advantages. These include the use of natural egg components, which avoids the use of viral vectors, small molecules, or chemical factors altogether. Moreover, ES cells derived via somatic cell NT, when compared to iPS cells, are able to be reprogrammed at higher efficiency and are of higher equality, as shown through having less epigenetic memory via each respective reprogramming step (Kim et al. 2010; Polo et al. 2010).

Recent headway made in human somatic cell NT – where non-enucleated human eggs were able to successfully reprogram diploid human somatic cells – suggests the utility of NT as a technique for generating pluripotent stem cells (Noggle et al. 2011). However, the aforementioned technique generated triploid cells, which would not be compatible with therapeutic application. Coupled with the ethical challenges associated with ES cell research and the scarcity of human embryos available for research purposes, these roadblocks provide a great challenge in the application of NT to human therapy (Egli et al. 2011). How can we transition from the above to develop useful somatic cell NTs?

NT to Enucleated Eggs

Originally established in amphibia, the initial NT experiments had a nucleus from a ruptured cell injected into an enucleated and unfertilized egg (Briggs and King 1952). A proportion of these were able to develop normally through embryogenesis, reaching adulthood (Gurdon et al. 1958). When the donor nuclei were taken from more embryonic cells, such as from a blastula, a higher proportion was able to reach blastula stages, along with adulthood (Gurdon 1960). Thus, in general, less terminally committed cells are less resistant to nuclear reprogramming by eggs. Furthermore, it has been demonstrated that transplantation of a mammalian somatic cell nucleus into an enucleated egg in second meiotic metaphase (MII) can lead to

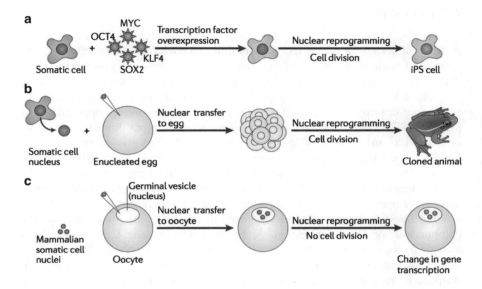

Fig. 1 Nuclear reprogramming methods. (**a**) Transcription factor overexpression. Oct4, Klf4, Sox2, and c-Myc expression reprograms somatic cells to an ES cell-like state, generating iPS cells. (**b**) Somatic cell NT. A somatic cell nucleus is transplanted into an enucleated egg in second meiotic methaphase, which allows nuclear reprogramming of the transplanted genome towards a pluripotent state. (**c**) When mammalian somatic cell nuclei are transplanted into the germinal vesicle (GV) of Xenopus laevis oocytes at first meiotic prophase, no cell division occurs. However, factors present in the oocyte directly reprogram gene expression (Jullien et al. 2011)

successful development of NT embryos (Campbell et al. 1996; Dominko et al. 1999). Various efficiencies of mammalian NT have been extensively reviewed (Beyhan et al. 2007).

Nuclear reprogramming via NT to MII oocytes can involve a high number of cell divisions with DNA synthesis. As such, inefficient reprogramming can result from the inability of transplanted nuclei to synchronize with these rapid cell cycle changes, leading to abnormal chromosomes and failure of further development (Mizutani et al. 2012). Still, *Oct4* reactivation during mouse somatic cell NT is able to occur 1–2 cell divisions post-NT, whereas derivation of germ cells requires an average 25 cell divisions post-NT (Boiani et al. 2005; Egli and Eggan 2006). Therefore, a system wherein cell division is not a compounding factor would be better for studying mechanisms of reprogramming in NT (Fig. 1).

NT to Oocytes

A different design for NT exists, wherein multiple nuclei are transplanted into the germinal vesicle (GV) of amphibian first meiotic prophase oocytes (Byrne et al. 2003). The oocyte GV contains a high concentration of components, eventually distributed to the rest of the egg post-meiotic maturation, that is necessary for embryonic development (Gao 2002). Of particular note, and as opposed to the previous NT system, this one does not generate new cell types. Somatic cell nuclei injected into an oocyte GV do not undergo DNA synthesis or cell division; however, they become intensely active in RNA synthesis along with the host oocyte. Thus, it is possible to transplant multiple nuclei from mammalian cells into the amphibian oocyte and see activation of genes that are active during normal early development, such as pluripotency genes. Furthermore, it is possible to observe direct activation of silenced genes in adult somatic nuclei without the complication of DNA replication. A direct switch in gene transcription from somatic to oocyte-type occurs without the intervention of or need for DNA replication. Thus, the exchange of factors involved directly reflects the process of transcriptional reprogramming. Conversely, as previously described, the NT experiments to unfertilized eggs are complicated by a period of intense DNA replication and cell division, along with the absence of transcription immediately following NT. As such, the timing of transcriptional reprogramming in egg NT experiments is difficult to analyze.

Using the NT to oocyte system, the oocyte-type linker histone B4 was identified as a necessary factor for efficient gene reactivation (Jullien et al. 2010). Following NT, B4 is incorporated into transplanted nuclei, a process that is associated with the loss of somatic linker histone H1. Furthermore, it was found that polymerization of nuclear actin, which is especially abundant in the oocyte GV, is necessary for transcriptional reactivation of *Oct4* in the oocyte system during reprogramming (Miyamoto et al. 2011).

It can be concluded that *Xenopus* oocytes are able to efficiently induce gene reactivation, without cell division and within a short window of time, utilizing natural oocyte components. Thus, the transplantation of multiple mammalian nuclei to *Xenopus* oocytes can be seen as a model system for investigating the mechanisms of transcriptional reprogramming. Additionally, their size (1.2 mm in diameter) allows for easier manipulation and an abundant source of material for identification of novel factors in reprogramming.

Potential Therapeutic Benefits of NT Reprogramming

The three main routes by which nuclear reprogramming can be achieved are induced pluripotency by transcription factor overexpression, cell fusion, and NT to eggs or oocytes. Cell fusion involves retention of one of the nuclear donors and does not, therefore, yield a cell whose genetic material is all derived from the intended donors. Transcription factor overexpression for induced pluripotency is an excellent procedure but the yield of reprogrammed cells is initially very small, which could present problems if large numbers of iPS cells are required.

Reprogramming by NT has some disadvantages and some advantages. The disadvantages are that NT to unfertilized eggs (in MII) is generally followed by extensive abnormalities of development when donor nuclei from differentiated cells are used. The yield of normal cells is, therefore, very small. NT to oocytes (in first meiotic prophase) does not yield new growing cells, although it does achieve pluripotency gene expression in transplanted nuclei. A further disadvantage of NT is that it is unlikely to be practical for humans because of the great difficulty in obtaining sufficient numbers of human eggs (Egli et al. 2011). There is also the problem that, so far, NT in humans has succeeded only when the egg nucleus is retained to make a fusion with an introduced somatic nucleus; until a procedure is developed by which the egg nucleus can be eliminated and NT achieved with only donor nuclear material, this also presents an obstacle to practical use.

The advantages of a NT route for reprogramming are that it makes use of natural components contained in eggs. It is important to remember that eggs are able to reprogram (or activate) pluripotency genes in the highly specialized and condensed sperm nuclei with 100 % efficiency. A sperm nucleus is more specialized than any somatic nucleus. The egg, therefore, possesses a remarkable supply of components that are able to achieve this activation of a sperm nucleus. It might well be that the natural gene-activating components of eggs could yield more perfectly reprogrammed somatic nuclei than enforced reactivation by overexpressed transcription factors. An advantage of using oocytes (as opposed to eggs) for reprogramming somatic nuclei is that it might well be possible, in future, to identify the actual components of oocytes or eggs that provide the reprogramming effect. This point is particularly true of amphibian oocytes, because 1 frog contains some 25,000 oocytes, each of which is over a millimeter in diameter. Therefore, the amount of material available for analysis is enormous compared to mammals, most of which have eggs of <100 µm and a limited number of these are available.

In conclusion, the NT route towards reprogramming is likely to be of eventual therapeutic value if oocyte or egg components can be identified, purified, and used alone or in conjunction with transcription factor overexpression to achieve large numbers of well-reprogrammed somatic cells.

References

Beyhan Z, Iager AE, Cibelli JB (2007) Interspecies nuclear transfer: implications for embryonic stem cell biology. Stem Cell 1:502–512

Boiani M, Gentile L, Gambles VV, Cavaleri F, Redi CA, Schöler HR (2005) Variable reprogramming of the pluripotent stem cell marker Oct4 in mouse clones: distinct developmental potentials in different culture environments. Stem Cells 23:1089–1104

Briggs R, King TJ (1952) Transplantation of living nuclei from blastula cells into enucleated frogs' eggs. Proc Natl Acad Sci USA 38:455–463

Byrne JA, Simonsson S, Western PS, Gurdon JB (2003) Nuclei of adult mammalian somatic cells are directly reprogrammed to oct-4 stem cell gene expression by amphibian oocytes. Curr Biol 13:1206–1213

Campbell KH, McWhir J, Ritchie WA, Wilmut I (1996) Sheep cloned by nuclear transfer from a cultured cell line. Nature 380:64–66

Chen J, Liu J, Chen Y, Yang J, Chen J, Liu H, Zhao X, Mo K, Song H, Guo L, Chu S, Wang D, Ding K, Pei D (2011) Rational optimization of reprogramming culture conditions for the generation of induced pluripotent stem cells with ultra-high efficiency and fast kinetics. Cell Res 21:884–894

Dominko T, Mitalipova M, Haley B, Beyhan Z, Memili E, McKusick B, First NL (1999) Bovine oocyte cytoplasm supports development of embryos produced by nuclear transfer of somatic cell nuclei from various mammalian species. Biol Reprod 60:1496–1502

Egli D, Eggan K (2006) Nuclear transfer into mouse oocytes. JoVE. doi:10.3791/116

Egli D, Chen AE, Saphier G, Powers D, Alper M, Katz K, Berger B, Goland R, Leibel RL, Melton DA, Eggan K (2011) Impracticality of egg donor recruitment in the absence of compensation. Cell Stem Cell 9:293–294

Gao S (2002) Germinal vesicle material is essential for nucleus remodeling after nuclear transfer. Biol Reprod 67:928–934

Gore A, Li Z, Fung HL, Young JE, Agarwal S, Antosiewicz-Bourget J, Canto I, Giorgetti A, Israel MA, Kiskinis E, Lee JH, Loh YH, Manos PD, Montserrat N, Panopoulos AD, Ruiz S, Wilbert ML, Yu J, Kirkness EF, Izpisua Belmonte JC, Rossi DJ, Thomson JA, Eggan K, Daley GQ, Goldstein LS, Zhang K (2011) Somatic coding mutations in human induced pluripotent stem cells. Nature 471:63–67

Gurdon JB (1960) The developmental capacity of nuclei taken from differentiating endoderm cells of *Xenopus laevis*. J Embryol Exp Morphol 8:505–526

Gurdon JB, Elsdale TR, Fischberg M (1958) Sexually mature individuals of *Xenopus laevis* from the transplantation of single somatic nuclei. Nature 182:64–65

Hussein SM, Batada NN, Vuoristo S, Ching RW, Autio R, Närvä E, Ng S, Sourour M, Hämäläinen R, Olsson C, Lundin K, Mikkola M, Trokovic R, Peitz M, Brüstle O, Bazett-Jones DP, Alitalo K, Lahesmaa R, Nagy A, Otonkoski T (2011) Copy number variation and selection during reprogramming to pluripotency. Nature 471:58–62

Jullien J, Astrand C, Halley-Stott RP, Garrett N, Gurdon JB (2010) Characterization of somatic cell nuclear reprogramming by oocytes in which a linker histone is required for pluripotency gene reactivation. Proc Natl Acad Sci USA 107:5483–5488

Jullien J, Pasque V, Halley-Stott RP, Miyamoto K, Gurdon JB (2011) Mechanisms of nuclear reprogramming by eggs and oocytes: a deterministic process? Nat Rev Mol Cell Biol 22:453–459

Kim K, Doi A, Wen B, Ng K, Zhao R, Cahan P, Kim J, Aryee MJ, Ji H, Ehrlich LI, Yabuuchi A, Takeuchi A, Cunniff KC, Hongguang H, McKinney-Freeman S, Naveiras O, Yoon TJ, Irizarry RA, Jung N, Seita J, Hanna J, Murakami P, Jaenisch R, Weissleder R, Orkin SH, Weissman IL, Feinberg AP, Daley GQ (2010) Epigenetic memory in induced pluripotent stem cells. Nature 467:285–290

Lister R, Pelizzola M, Kida YS, Hawkins RD, Nery JR, Hon G, Antosiewicz-Bourget J, O'Malley R, Castanon R, Klugman S, Downes M, Yu R, Stewart R, Ren B, Thomson JA, Evans RM, Ecker JR (2011) Hotspots of aberrant epigenomic reprogramming in human induced pluripotent stem cells. Nature 471:68–73

Miyamoto K, Pasque V, Jullien J, Gurdon JB (2011) Nuclear actin polymerization is required for transcriptional reprogramming of Oct4 by oocytes. Genes Dev 25:946–958

Mizutani E, Yamagata K, Ono T, Akagi S, Geshi M, Wakayama T (2012) Abnormal chromosome segregation at early cleavage is a major cause of the full-term developmental failure of mouse clones. Dev Biol 364:56–65

Noggle S, Fung HL, Gore A, Martinez H, Satriani KC, Prosser R, Oum K, Paull D, Druckenmiller S, Freeby M, Greenberg E, Zhang K, Goland R, Sauer MV, Leibel RL, Egli D (2011) Human oocytes reprogram somatic cells to a pluripotent state. Nature 478:7075

Okita K, Matsumura Y, Sato Y, Okada A, Morizane A, Okamoto S, Hong H, Nakagawa M, Tanabe K, Tezuka K, Shibata T, Kunisada T, Takahashi M, Takahashi J, Saji H, Yamanaka S (2011) A more efficient method to generate integration-free human iPS cells. Nat Meth 8:409–412

Polo JM, Liu S, Figueroa ME, Kulalert W, Eminli S, Tan KY, Apostolou E, Stadtfeld M, Li Y, Shioda T, Natesan S, Wagers AJ, Melnick A, Evans T, Hochedlinger K (2010) Cell type of origin influences the molecular and functional properties of mouse induced pluripotent stem cells. Nat Biotechnol 28:848–855

Takahashi K, Yamanaka S (2006) Induction of pluripotent stem cells from mouse embryonic and adult fibroblast cultures by defined factors. Cell 126:663–676

Wang W, Yang J, Liu H, Lu D, Chen X, Zenonos Z, Campos LS, Rad R, Guo G, Zhang S, Bradley A, Liu P (2011) Rapid and efficient reprogramming of somatic cells to induced pluripotent stem cells by retinoic acid receptor gamma and liver receptor homolog 1. Proc Natl Acad Sci USA 45:18283–18288

Young MA, Larson DE, Sun CW, George DR, Ding L, Miller CA, Lin L, Pawlik KM, Chen K, Fan X, Schmidt H, Kalicki-Veizer J, Cook LL, Swift GW, Demeter RT, Wendl MC, Sands MS, Mardis ER, Wilson RK, Townes TM, Ley TJ (2012) Background mutations in parental cells account for most of the genetic heterogeneity of induced pluripotent stem cells. Cell Stem Cell 10:570582

Zhang G, Shang B, Yang P, Cao Z, Pan Y, Zhou Q (2012) Induced pluripotent stem cell consensus genes: implication for the risk of tumorigenesis and cancers in induced pluripotent stem cell therapy. Stem Cells Dev 21:955–964

Induction of Neural Lineages from Mesoderm and Endoderm by Defined Transcription Factors

Marius Wernig

Abstract A major interest in developmental biology is the lineage plasticity of specialized cells. We recently generated induced neuronal (iN) cells from fibroblasts and hepatocytes by expression of defined pan-neuronal transcription factors. Moreover, we were able to generate induced neural precursor (iNP) cells following a similar transcription factor-based strategy, suggesting that cell fate plasticity is much wider than previously anticipated. Further studies showed that addition of subtype-specific regulators was sufficient to induce characteristic neuronal subtypes such as dopamine and motor neurons. Here, we review the most recent developments of our own research efforts as well as the relevant findings in the field as they pertain to direct induction of neural lineages. The derivation of neuronal cells from patient fibroblasts holds great promise to uncover human neurological disease mechanisms and to provide a donor source for autologous therapeutic transplantation. Similar to induced pluripotent stem (iPS) cells, much has yet to be learned about iN cells before clinical translation can be realized.

Introduction

The induction of pluripotency in somatic cells by defined factors or somatic cell nuclear transfer provided unambiguous evidence that the epigenetic state of terminally differentiated somatic cells is not static and can be reversed to a more primitive state (Gurdon 2006; Jaenisch and Young 2008; Yamanaka 2007). Inspired by these results, we have recently identified approaches to directly convert fibroblasts into induced neuronal (iN) cells, indicating that direct lineage conversions are possible between very distantly related cell types (Vierbuchen et al. 2010). Importantly,

M. Wernig (✉)
Institute for Stem Cell Biology and Regenerative Medicine, Department of Pathology
Stanford University School of Medicine, 265 Campus Drive, Stanford, CA 94305, USA
e-mail: wernig@stanford.edu

F.H. Gage and Y. Christen (eds.), *Programmed Cells from Basic Neuroscience to Therapy*, Research and Perspectives in Neurosciences 20,
DOI 10.1007/978-3-642-36648-2_3, © Springer-Verlag Berlin Heidelberg 2013

iN cells could also be derived from defined endodermal cells and not only induced neuronal properties but also extinguished their donor cell identity. Therefore, this experimental lineage reprogramming represented a complete and functional lineage switch as opposed to chimeric phenotypes. Since the discovery in 1987 that the bHLH class transcription factor MyoD can induce a myogenic program in fibroblast cells, several other examples of remarkable cell-fate changes have been observed in response to forced expression of transcriptional regulators, but until recently it was assumed that this phenomenon was limited to closely related cell lineages (Graf and Enver 2009; Zhou and Melton 2008). Following the description of iN cells, it was demonstrated that fibroblasts could be directly converted to a diverse range of cell types, such as cardiomyocytes (Ieda et al. 2010), blood cell progenitors (Szabo et al. 2010), and hepatocytes (Huang et al. 2011; Sekiya and Suzuki 2011). Here, I discuss recent achievements in direct lineage reprogramming towards the neuronal lineage (Fig. 1).

Direct Reprogramming of Mouse Fibroblasts to Neurons

Given the determinant role of transcription factors in cell fate specification, we hypothesized that forced expression of a combination of such factors might be sufficient to directly convert mouse fibroblasts into neuronal cells (Vierbuchen et al. 2010). Initially, we expressed a pool of 19 candidate transcription factors in cultured mouse fibroblasts. Surprisingly, after 4 weeks in culture, some cells exhibited a very typical neuronal morphology, with complex arborization of their processes and expressed the pan-neuronal markers Tau and Tuj1. We then attempted to define the minimal pool of genes required for this conversion. Omission of any of the 19 factors did not yield conclusive results. However, when testing individual factors, we found that the bHLH transcription factor Ascl1 was sufficient to induce immature neuronal morphologies. A screen of the other 18 factors in two-factor-combinations with Ascl1 demonstrated that the additional four genes – Brn2/4, Zic1, Mytl1, and Olig2 – greatly facilitated the effect of Ascl1 and, when combined, yielded cells with complex, mature neuronal morphologies. Of those five factors, we then found that just three – Ascl1, Brn2, and Mytl1 – are sufficient to induce neuronal cells. Importantly, the reprogrammed cells not only expressed a broad spectrum of pan-neuronal markers but also were able to generate action potentials and form synapses, the two principal properties of neurons. The establishment of complex functional properties such as pre- and postsynaptic compartments indicated that an array of neuronal genes was induced and that those gene products assembled into well-coordinated functional units. We therefore termed the cells iN cells.

Surprisingly, the conversion efficiency of embryonic fibroblasts was estimated to be close to 20 % within 2 weeks, indicating that the conversion toward neuronal fates is faster and more efficient than induced pluripotent stem (iPS) cell formation. Also in contrast to induction of pluripotency, iN cell reprogramming does not

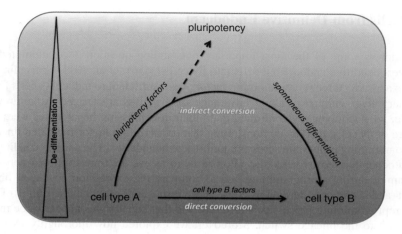

Fig. 1 Different modes of lineage reprogramming. Two principle strategies have been suggested that allow the conversion of one somatic cell type into another. One approach, that I would like to call "direct conversion," utilizes lineage-determining transcription factors normally expressed in the target cell type (straight arrow from cell type A to cell type B). Another approach – I suggest the term "indirect conversion" – utilizes transcription factors that are known to induce pluripotency. In contrast to iPS cell reprogramming, the reprogramming factors are only induced for a short period of time that is not sufficient to reach full pluripotency but presumably some partially reprogrammed, unstable pluripotent state from which any desired cell type can be generated through spontaneous differentiation. Exposure to selective media (developed for guided iPS cell differentiation) can enrich for a desired cell type

require DNA synthesis, as judged by the lack of BrdU incorporation during the process. Thus, the genome-wide epigenetic remodeling must happen without cell division, which is believed to be a key mediator of iPS cell reprogramming (Hong et al. 2009; Kawamura et al. 2009; Li et al. 2009; Marion et al. 2009; Utikal et al. 2009). Cell proliferation is arguably a more favorable condition for epigenetic changes; that is, however, at odds with the contrasting low iPS cell reprogramming efficiencies of about 1 %.

Our study raised many new questions that we and others have now begun to address. Among other things, we felt it was important to determine (1) what the exact cell of origin for iN cells was, (2) whether endodermal cells could be coaxed towards iN cells, (3) whether the reprogramming factors initiated the neurogenic program while suppressing the original cell fates or iN cells retained molecular reminiscences of their cell of origin, (4) whether the reprogramming involved an intermediate neural progenitor cell, (5) whether iN cells possessed a regional identity, (6) whether modifying the combination of transcription factors would bestow a specific neural subtype identity, (7) whether iN cells could functionally integrate into the brain, and last but not least, (8) whether methods could be developed to generate human iN cells. Recently, we and others have provided some answers to these fundamental questions and I will discuss them in the following sections.

iN Cells from Definitive Endoderm

To address some of the most important outstanding questions, we attempted to convert definitive endodermal cells into iN cells (Marro et al. 2011). Intriguingly, the exact same three reprogramming factors were sufficient to induce iN cells from primary liver cultures. Taking advantage of a well-characterized Albumin-Cre allele, we unequivocally confirmed that albumin-expressing hepatocytes were the origin of iN cells, thereby demonstrating that transcription factor-mediated lineage reprogramming is possible across major lineage boundaries. The genetic labeling of hepatocyte-derived iN cells also allowed us to assess the transcriptional network dynamics during reprogramming and to compare the expression profile of fibroblast- and hepatocyte-derived iN cells. These results indicated that the timing of the two reprogramming processes is different and that hepatocytes appear to be more resistant to the lineage switch. Moreover, we explored how thoroughly iN cells were reprogrammed. We found that the donor cell type-specific expression signatures were robustly silenced in both fibroblast- and hepatocyte-derived iN cells, which led us to the somewhat surprising conclusion that exactly the same three transcription factors were able to not only induce a neuronal program in different types of cells of origin but also to downregulate two completely unrelated original transcriptional programs. Detailed gene expression analysis at the population and single-cell level indicated that iN cells possessed a small degree of epigenetic memory of their donor cells and that these transcriptional remnants decreased over time. It will be interesting to investigate the molecular mechanisms underlying the transcriptional silencing, which could also be utilized during cell-fate specification in the embryo. Given the substantial differences in the fibroblast and liver transcription programs, it seems unlikely that the transcriptional silencing is directly mediated by the neuronal transcription factors. However, it is a formal possibility that the BAM factors target and inhibit a large number of key lineage-determining factors representing many non-neuronal cell fates. Alternatively, the mutual lineage switch could be caused by a more general mechanism. Perhaps when cells are becoming specified to one particular lineage, a process becomes activated that leads to transcriptional silencing of other lineage programs. For example, lineage-determining factors may have to compete for a finite amount of ubiquitously expressed and required co-factors, which would lead to an obligatory extinction of undesired lineages once differentiating cells have committed to one lineage. The ubiquitously expressed E-proteins could represent such critical co-factors, as they are known to heterodimerize with several different, lineage-specific bHLH transcription factors such as MyoD, Ascl1 and others (Massari and Murre 2000).

Can Human Fibroblasts Be Converted to Neuronal Cells?

Another important question that remained unclear was whether iN cells could also be generated from human fibroblasts. This is a significant question because potential clinical applications could only be realized with human cells. Given that the exact same four transcription factors can reprogram fibroblasts from both mouse and human into iPS cells, one would expect that converting human fibroblasts to iN cells can be achieved with very similar, if not identical, methods as mouse fibroblasts. However, when the BAM factors were introduced into human fetal fibroblasts, the resulting cells remained immature and failed to generate action potentials when depolarized (Pang et al. 2011). We therefore screened 20 additional factors, including more transcription factors but also sets of microRNAs and chromatin factors, in combination with BAM, and we found that, by introducing the bHLH factor NeuroD1, functional neurons from human fibroblasts could be generated. The human iN cells expressed a variety of neuronal markers, including Tuj1, MAP2, NeuN, neurofilaments and synapsins, and exhibited functional neuronal properties as judged by measurement of action potentials. Importantly, when cultured with primary cortical neurons, both spontaneous and evoked postsynaptic currents could be detected in these cells, demonstrating their synaptic maturation. However, the first functional synapses were found only after 5–6 weeks, suggesting that full maturation of human iN cells is a slow process. We then went on to show that the same four factors were shown to also convert postnatal foreskin fibroblasts into synaptically competent iN cells with comparable timing and efficiency. Similar to mouse iN cells, the majority of human iN cells expressed mRNAs characteristic of glutamatergic neurons, such as vGLUT1 and vGLUT2. After downregulation of the exogenous transcription factors, the iN cells retained their stability, which indicated that an intrinsic program was established to maintain the newly adopted neuronal identity. The overall efficiency to generate human iN cells with four factors (2–4%) was about tenfold lower than that of mouse iN cells with just three factors (compare to Vierbuchen et al. 2010). The observed species differences in iN cell reprogramming may appear unexpected in light of the robustness of generating human iPS cells. However, upon closer inspection, the two different human reprogramming paradigms do share many similarities. The drop in human iPS cell reprogramming efficiencies is of similar scale, especially when taking into account the fact that only a small fraction of plated fibroblasts and only a small subset of these cells' progenies are forming iPS cells. Similarly, it takes much longer until iPS cell-like colonies appear in human versus mouse fibroblasts. Thus, human cells in general appear to be more rigid and to possess a higher epigenetic hurdle to both iN and iPS cell reprogramming. Finally, human embryonic stem (ES) cell-derived neurons require similar amounts of time to develop synaptic competence as human iN cells (Johnson et al. 2007; Wu et al. 2007). Thus, a long maturation period may be an inherent property of human cells, which is perhaps not surprising given that human brain development is orders of magnitude slower than rodent brain development.

In attempts to convert adult human fibroblasts to neurons, another group resorted to the combination of those five factors that were initially found in mouse to be the most critical of the 19 tested candidate factors. The resulting cells possessed a series of neuronal properties, such as the ability to generate action potentials when depolarized (Qiang et al. 2011). The acquisition of more mature functional properties, such as synaptic transmission, was less clear. This observation is similar to ours when adult fibroblasts were infected with BAM and Neurod1 (Pang et al. 2011). Nevertheless, iN cells derived from familial Alzheimer's disease (FAD) patients with mutations in *PRESENILIN* genes were found to potentially exhibit disease-specific traits, demonstrating that iN cells can be used to model human disease. The FAD-iN cells showed the presence of amyloid precursor protein (APP) puncta in endosomes, not readily detected in the originating FAD fibroblasts (Qiang et al. 2011). This phenotype could be rescued by overexpression of wild-type PRESENILIN1. Although a more detailed analysis of phenotypic differences between wild-type and FAD-iN cells will have to be performed in the future, this study underscored the potential utility of iN cell technology in the study of neurodegenerative diseases. Of note, one of the reprogramming factors used in this study is Olig2, another member of bHLH family. Olig2 is not specific to neurons and was shown to promote both neuronal and oligodendroglial fates, depending on the developmental context (Lu et al. 2002; Mizuguchi et al. 2001; Novitch et al. 2001). However, in contrast to Ascl1 and NeuroD1, Olig2 is thought to act as a repressor and has been shown to associate with Ngn2 and E47 to antagonize their neurogenic effect in the context of motor neuron formation (Lee et al. 2005). Future work will have to elucidate the function of these transcription factors during reprogramming. Given that different transcription factor combinations can induce neuronal cells, there may be several parallel ways towards the neuronal lineage. Alternatively, the different transcription factors may eventually activate the same core program to induce neuronal identity.

It is not surprising that the vast majority of reprogramming factors known to date are transcriptional regulators, given their ability to directly activate gene expression and the existence of "master regulators" for specific lineages such as the muscle lineage (Weintraub et al. 1989). It is surprising, however, that miRNAs, which are believed to predominantly function through downregulation of gene activity, seem to be powerful agents in reprogramming. Recently, Yoo et al. (2011) showed that by introducing miR-9/9* and miR-124, human fibroblasts can be reprogrammed into cells with neuron-like morphologies expressing the pan-neuronal marker MAP2. While these phenotypic changes are truly remarkable, the miRNAs alone were not sufficient to induce functional iN cells. However, the addition of the transcription factors NEUROD2, ASCL1, and MYT1L greatly increased the conversion efficiencies and led to the formation of iN cells from fetal and adult human fibroblasts with all principal functional properties of neurons, including synapse formation. Intriguingly, this report also underscored the essential role of bHLH transcription factors for generation of human iN cells. miR-9* and miR-124 are specifically expressed in post-mitotic neurons and were shown to repress the expression of SWI/SNF complex subunit Baf53a. When neural progenitor cells exit the cell cycle and differentiate into neurons, Baf53a is replaced by Baf53b and this switch is functionally relevant in vivo (Yoo et al. 2009). Therefore, it appeared likely that the miRNAs facilitated reprogramming through promoting this BAF complex subunits switch. However, prolonging the expression of BAF53a did not abolish the conversion from fibroblasts to neurons and, therefore, downregulation of this miRNA target does not seem to be critical in this context (Yoo et al. 2011). Since we showed iN cell induction by Neurod1, Ascl1, Myt1l, and Brn2 alone without miRNAs, it appears that the miRNAs are able to replace the transcription factor Brn2 (Pang et al. 2011). This was not tested directly, however, and it is possible that the miRNAs work through yet another mechanism.

More recently, another group also found miR-124 to be beneficial for human iN cell formation (Ambasudhan et al. 2011). Interestingly, in that report the miRNA was combined with BRN2 and MYT1L, suggesting that the miRNAs do have a complementary function to Brn2. Surprisingly, no bHLH transcription factors were used in the latter report, but the miR-124/Brn2/Myt1l-iN cells appeared to be less faithfully reprogrammed, judging by the lack of convincing evidence of synaptic competence. Future studies on the miRNA-transcription factor interplay responsible for iN cell formation could also be relevant to regular neural development. Thus, the extremely non-physiological method of reprogramming could become a discovery tool for studying normal development.

Induction of Specific Neuronal Subtypes from Fibroblasts

For clinical and experimental use of iN cells, it would be desirable to develop ways to generate neurons with neurotransmitter- and region-specific phenotypes. Liver- and fibroblast-iN cells generated with the same reprogramming factors displayed properties characteristic of excitatory neurons, suggesting that the glutamatergic fate is a default fate, as has been suggested for ES cell differentiation systems (Gaspard et al. 2008). Alternatively, the choice of transcription factors could have specifically induced an excitatory subtype. Therefore, the question arises whether inclusion of subtype-specific transcription factors in the reprogramming factor combination may direct cells into other desired subtypes.

This hypothesis was elegantly tested by Son et al. (2011), who attempted to generate iN cells with motor neuron identity directly from fibroblasts. They started out with a fairly large pool of transcription factors critical for motor neuron specification, and eventually found four of them (Lhx3, Hb9, Isl1 and Ngn2) that, in combination with the BAM factors, generated Hb9-positive neurons with efficiencies up to 10 % from MEFs (mouse embryonic fibroblasts). Gene expression analyses indicated that these induced motor neuronal (iMN) cells resembled the transcription profiles of embryonic and ES cell-derived motor neurons. Besides displaying electrophysiological properties akin to motor neurons, these iMN cells formed functional synaptic connections with myotubes. When transplanted into the developing chick spinal cord, most of these iMN cells were engrafted in the ventral horn of the spinal cord with axons projecting into the ventral roots. In addition, the cells behaved similarly to ES cell-derived motor neurons in disease conditions. When cultured with glia carrying the G93A mutation in the *Superoxide dismutase* (*Sod1*) gene, a mutation found in familial forms of amyotrophic lateral sclerosis (ALS), the survival of iMN cells decreased. Vice versa, iMN cells derived from *Sod1G93A* MEFs also showed reduced survival when cultured with wild-type glia. These first translational studies suggest that iMN cells can be used as a tool to understand the pathophysiology of ALS. As a first step into this direction, human ES cell-derived, fibroblast-like cells were infected with the seven transcription factors in combination with Neurod1. This approach yielded neuronal cells that could fire action potentials and expressed Hb9 and vesicular ChAT. More work is needed to investigate whether iMN cells can be generated from primary human fibroblasts.

Another clinically relevant neuronal subtype that has been under intense investigation is the group of midbrain dopaminergic (DA) neurons, which are preferentially affected in Parkinson's disease. Recently, two important proof-of-principle studies described the generation of iN cells expressing tyrosine hydroxylase (TH), the rate-limiting enzyme in catecholamine biosynthesis (Caiazzo et al. 2011; Pfisterer et al. 2011). Pfisterer et al. identified Lmx1a and Foxa2, when used in conjunction with the BAM pool, to be capable of generating iN cells expressing TH, Aromatic L-amino acid decarboxylase (AADC), which is another crucial enzyme in catecholamine biosynthesis, and more importantly, Nurr1, a marker of midbrain identity. However, the cells did not express other midbrain markers and were not able to release dopamine into the media. Another report (Caiazzo et al. 2011) demonstrated the generation of mouse iN cells with dopaminergic features by expression of the transcription factors Ascl1, Nurr1 and Lmx1a. The efficiency of induction of TH-positive cells was reported to be 18 % relying on a TH-EGFP transgenic reporter line. The TH-positive cells co-expressed vesicular monoamine transporter 2 (VMAT2), dopamine transporter (DAT), aldehyde dehydrogenase 1a1 (ALDH1A1), and calbindin. In contrast to the BAM/Foxa2/Lmx1a cells described by Pfisterer et al., the Ascl1/Nurr1/Lmx1a iN cells were able to release dopamine as determined by amperometry and HPLC analysis, which indicated that the cells exhibited an important functional property of dopamine neurons. Intriguingly, similar results could be obtained using human fibroblasts and the same reprogramming factors. However, the resulting cells did not express any regional markers specific to midbrain and displayed immature morphologies. Moreover, it was not investigated whether these cells were competent to receive synaptic inputs. Therefore, despite the use of midbrain dopamine neuron-specific transcription factors for reprogramming, only generic dopamine neurons and no midbrain-specific features were observed, suggesting incomplete reprogramming. Genome-wide transcriptional profiling showed that the induced dopamine neuron-like cells were more similar to brain-derived dopamine neurons than fibroblasts, but this may be true for any neuronal subtype and many genes were still differentially expressed.

As a cautionary note, the lack of midbrain character is a critical limitation for clinical application, since only "authentic" human midbrain dopamine neurons are able to restore function in animal models of Parkinson's disease (Kriks et al. 2011; Roy et al. 2006; Yang et al. 2008). Therefore, yet another group very recently attempted to generate iN cells that are more reminiscent of midbrain dopamine neurons (Kim et al. 2011b). This time, the transcription factor combinations were screened to induce EGFP fluorescence in Pitx3::EGFP knock-in fibroblasts, a locus highly specific for midbrain dopamine neurons. Surprisingly, EGFP+ cells were readily detected with a combination of two factors (Ascl1 and Pitx3). Complementation with another four factors (Nurr1, Lmx1a, Foxa2, and En1) as well as the patterning factors Shh and FGF8 further enhanced the amount of EGFP+ cells. The EGFP+ cells also expressed the generic dopamine neurons markers TH, DAT, AADC, and VMAT2 and were able to release dopamine. However, when tested in vivo, the cells only partially restored dopamine function and, when a series of midbrain markers were analyzed, both the two-factor and the six-factor iN cells failed to reach transcription levels similar to those found in embryonic or adult midbrain dopamine neurons. This finding leads to the somewhat sobering conclusion that even the combination of six transcription factors may still not be sufficient to fully reprogram fibroblasts to this specific neuronal subtype.

Conversion of Fibroblasts into Neural Precursor Cells

Directly generating terminally differentiated neurons can be useful in disease modeling and transplantation studies. However, a clear limitation of postmitotic iN cells is their inability to expand once reprogrammed. Large numbers will be required for cell replacement-based therapies in a clinical setting. Therefore, it would be desirable to induce expandable neural precursor cells directly from fibroblasts. Recently, Ding and colleagues were successful in converting mouse fibroblasts to induced neural precursor (iNP) cells (Kim et al. 2011a). Surprisingly, in this study, the recipe for reprogramming was not tailored to the target cell type but was identical to the iPS cell reprogramming factors. In contrast to iPS cell formation, the factors were induced only for a short time followed by exposure to media favoring the growth of neural progenitor cells. After some optimization of the timing and culture media, colonies appeared that closely resembled, and expressed several markers of, mitotic neural rosette cells. Upon spontaneous differentiation, iNP cells could give rise to multiple neuronal subtypes and astrocytic cells, but cells with oligodendrocytic characteristics were not found, indicating that the iNP cells were at least bi-potent neural precursor cells. Very recently, another group confirmed that short exposure to the Yamanaka factors is sufficient to generate such iNP cells (Thier et al. 2012). It was clearly demonstrated that these iNP cells could also self-renew and differentiate into oligodendrocyte-like cells, suggesting their tri-lineage potential.

Remarkably, a very similar approach was taken to generate cardiomyocytes and blood progenitor cells from fibroblasts (Efe et al. 2011; Szabo et al. 2010). The authors suggested that the observed transdifferentiation bypassed an intermediate pluripotent stage because no Oct4 transcripts were observed and the short reprogramming factor expression was not compatible with iPS cell generation. However, given that the same approach can generate multiple somatic as well as pluripotent lineages, the simplest explanation is that the short-term expression of the pluripotency reprogramming factors indeed induces a transient pluripotent state, yet this state is unstable, prone to differentiate and cannot be stabilized by environmental cues only. While this approach is intriguing, the efficiencies of forming rosette-like colonies appeared low. In parallel studies we have attempted to directly induce neural precursor-like cells using neural rather than pluripotency transcription factors (Lujan et al. 2012). Of 11 transcription factors initially screened, we identified the 3 – Sox2, FoxG1, and Brn2 – that can in combination induce many features of proliferating neural precursor cells in mouse fibroblasts. Notably, the cells could self-renew and expand for many passages and differentiate into cells expressing neuronal, astrocytic and oligodendrocytic markers and morphologies. It appeared that Sox2 and FoxG1 was sufficient to induce populations of neuronal-restricted and neuronal/astroglial-restricted progenitor cells whereas the addition of Brn2 conferred a very efficient oligodendrocyte-differentiation capacity on the iNP cells. Importantly, those oligodendrocytes expressed a variety of markers and showed typical oligodendrocyte morphologies in vitro. After transplantation into the *Shiverer* mouse brain, which lacks large portions of the *myelin basic protein* (*Mbp*) gene, the cells again adopted typical morphologies, and immunostaining for Mbp demonstrated the cells' capacity to ensheathe host axons. Very recently, another report confirmed that mouse fibroblasts can be reprogrammed into iNP cells without using the key reprogramming factor Oct4 (Han et al. 2012). This group arrived at the four factors: Brn4, Sox2, Klf4 and c-Myc. Given the close functional and sequence similarity between Brn2 and Brn4 (Vierbuchen et al. 2010), it appears that FoxG1 may be replaceable by the combination of Klf4 and c-Myc. However, the natures of those transcription factors are fundamentally different, with FoxG1 being a largely forebrain-specific gene with lineage-specifying functions and Klf4 and c-Myc more broadly expressed and prominently involved in cell cycle regulation. An interesting speculation is whether a strong cell proliferation stimulus by Klf4 and c-Myc might enhance the effects of fewer lineage-specific transcription factors whereas perhaps fewer cell divisions might be required when more lineage-determining factors are ectopically expressed.

In any case, there are now two principally different but both transcription factor-mediated approaches to generating iNP cells. It remains to be seen what the communalities and differences of these two classes of iNP cells are. Much more work will need to be done to clarify the full functionality of the mature neural cell types derived from these cells (such as neurons and oligodendrocytes) in vitro and after transplantation and, importantly, whether similar cells can be generated from human fibroblasts.

Summary

While still in a nascent stage, the field of direct somatic lineage reprogramming into neural cell types has already attracted a lot of attention. From a biological stand-point, it may become a new method in the toolbox of developmental and molecular biology. It offers a new way to interrogate transcription factor function independent of the physiological environment and to study the complex interplay between sequence-specific transcriptional regulators and various repressive and active chro-matin states as well as the recruitment of their underlying chromatin-modifying enzymes. Moreover, the generation of iN cells represents a novel way to study the mechanisms of cell fate decisions of neural development and postmitotic neuronal maturation. The use of human iN cells provides an avenue to study human devel-opmental processes in live cultures, which may allow the discovery of species-specific differences compared to the much better studied model organisms.

From a medical point of view, direct lineage reprogramming provides an alternative, potentially complementary tool to many of the proposed applications of iPS cell technology for both disease modeling and development of cell-based therapies. The first reports of assessing disease-related phenotypes in iPS cell-derived neurons have just been published and provide an important proof-of-principle that at least some cellular aspects of complex brain diseases can be recapitulated with patient-derived cells in vitro (Marchetto et al. 2010; Brennand et al. 2011; Nguyen et al. 2011). While it is obviously much easier to generate large numbers of neuronal cells from iPS cells compared to direct conversion, a more and more appreciated limitation of iPS cells is line-to-line variability, complicating the discovery of subtle defects. Future studies will show whether directly generated iN cells may display a better representation of the cellular variability, which may simplify finding disease-associated phenotypes. Moreover, the generation of iN cells from a large cohort of patients appears quite feasible whereas the generation and neuronal differentiation of iPS cells would be a very cumbersome and slow process.

References

Ambasudhan R, Talantova M, Coleman R, Yuan X, Zhu S, Lipton SA, Ding S (2011) Direct reprogramming of adult human fibroblasts to functional neurons under defined conditions. Cell Stem Cell 9:113–118

Brennand KJ, Simone A, Jou J, Gelboin-Burkhart C, Tran N, Sangar S, Li Y, Mu Y, Chen G, Yu D, McCarthy S, Sebat J, Gage FH (2011) Modelling schizophrenia using human induced pluripotent stem cells. Nature 473:221–225

Caiazzo M, Dell'Anno MT, Dvoretskova E, Lazarevic D, Taverna S, Leo D, Sotnikova TD, Menegon A, Roncaglia P, Colciago G, Russo G, Carninci P, Pezzoli G, Gainetdinov RR, Gustincich S, Dityatev A, Broccoli V (2011) Direct generation of functional dopaminergic neurons from mouse and human fibroblasts. Nature 476:224–227

Efe JA, Hilcove S, Kim J, Zhou H, Ouyang K, Wang G, Chen J, Ding S (2011) Conversion of mouse fibroblasts into cardiomyocytes using a direct reprogramming strategy. Nature Cell Biol 13:215–222

Gaspard N, Bouschet T, Hourez R, Dimidschstein J, Naeije G, van den Ameele J, Espuny-Camacho I, Herpoel A, Passante L, Schiffmann SN, Gaillard A, Vanderhaeghen P (2008) An intrinsic mechanism of corticogenesis from embryonic stem cells. Nature 455:351–357

Graf T, Enver T (2009) Forcing cells to change lineages. Nature 462:587–594

Gurdon JB (2006) From nuclear transfer to nuclear reprogramming: the reversal of cell differentiation. Annu Rev Cell Dev Biol 22:1–22

Han DW, Tapia N, Hermann A, Hemmer K, Hoing S, Arauzo-Bravo MJ, Zaehres H, Wu G, Frank S, Moritz S, Greber B, Yang JH, Lee HT, Schwamborn JC, Storch A, Schöler HR (2012) Direct reprogramming of fibroblasts into neural stem cells by defined factors. Cell Stem Cell 10:465–472

Hong H, Takahashi K, Ichisaka T, Aoi T, Kanagawa O, Nakagawa M, Okita K, Yamanaka S (2009) Suppression of induced pluripotent stem cell generation by the p53–p21 pathway. Nature 460:1132–1135

Huang P, He Z, Ji S, Sun H, Xiang D, Liu C, Hu Y, Wang X, Hui L (2011) Induction of functional hepatocyte-like cells from mouse fibroblasts by defined factors. Nature 475:386–389

Ieda M, Fu JD, Delgado-Olguin P, Vedantham V, Hayashi Y, Bruneau BG, Srivastava D (2010) Direct reprogramming of fibroblasts into functional cardiomyocytes by defined factors. Cell 142:375–386

Jaenisch R, Young R (2008) Stem cells, the molecular circuitry of pluripotency and nuclear reprogramming. Cell 132:567–582

Johnson MA, Weick JP, Pearce RA, Zhang SC (2007) Functional neural development from human embryonic stem cells: accelerated synaptic activity via astrocyte coculture. J Neurosci 27:3069–3077

Kawamura T, Suzuki J, Wang YV, Menendez S, Morera LB, Raya A, Wahl GM, Izpisua Belmonte JC (2009) Linking the p53 tumour suppressor pathway to somatic cell reprogramming. Nature 460:1140–1144

Kim J, Efe JA, Zhu S, Talantova M, Yuan X, Wang S, Lipton SA, Zhang K, Ding S (2011a) Direct reprogramming of mouse fibroblasts to neural progenitors. Proc Natl Acad Sci USA 108:7838–7843

Kim J, Su SC, Wang H, Cheng AW, Cassady JP, Lodato MA, Lengner CJ, Chung CY, Dawlaty MM, Tsai LH, Jaenisch R (2011b) Functional integration of dopaminergic neurons directly converted from mouse fibroblasts. Cell Stem Cell 9:413–419

Kriks S, Shim JW, Piao J, Ganat YM, Wakeman DR, Xie Z, Carrillo-Reid L, Auyeung G, Antonacci C, Buch A, Yang L, Beal MF, Surmeier DJ, Kordower JH, Tabar V, Studer L (2011) Dopamine neurons derived from human ES cells efficiently engraft in animal models of Parkinson's disease. Nature 480:547–551

Lee SK, Lee B, Ruiz EC, Pfaff SL (2005) Olig2 and Ngn2 function in opposition to modulate gene expression in motor neuron progenitor cells. Genes Devel 19:282–294

Li H, Collado M, Villasante A, Strati K, Ortega S, Canamero M, Blasco MA, Serrano M (2009) The Ink4/Arf locus is a barrier for iPS cell reprogramming. Nature 460:1136–1139

Lu QR, Sun T, Zhu Z, Ma N, Garcia M, Stiles CD, Rowitch DH (2002) Common developmental requirement for Olig function indicates a motor neuron/oligodendrocyte connection. Cell 109:75–86

Lujan E, Chanda S, Ahlenius H, Sudhof TC, Wernig M (2012) Direct conversion of mouse fibroblasts to self-renewing, tripotent neural precursor cells. Proc Natl Acad Sci USA 109:2527–2532

Marchetto MC, Carromeu C, Acab A, Yu D, Yeo GW, Mu Y, Chen G, Gage FH, Muotri AR (2010) A model for neural development and treatment of Rett syndrome using human induced pluripotent stem cells. Cell 143:527–39

Marion RM, Strati K, Li H, Murga M, Blanco R, Ortega S, Fernandez-Capetillo O, Serrano M, Blasco MA (2009) A p53-mediated DNA damage response limits reprogramming to ensure iPS cell genomic integrity. Nature 460:1149–1153

Marro S, Pang ZP, Yang N, Tsai MC, Qu K, Chang HY, Sudhof TC, Wernig M (2011) Direct lineage conversion of terminally differentiated hepatocytes to functional neurons. Cell Stem Cell 9:374–382

Massari ME, Murre C (2000) Helix-loop-helix proteins: regulators of transcription in eucaryotic organisms. Mol Cell Biol 20:429–440

Mizuguchi R, Sugimori M, Takebayashi H, Kosako H, Nagao M, Yoshida S, Nabeshima Y, Shimamura K, Nakafuku M (2001) Combinatorial roles of olig2 and neurogenin2 in the coordinated induction of pan-neuronal and subtype-specific properties of motoneurons. Neuron 31:757–771

Nguyen HN, Byers B, Cord B, Shcheglovitov A, Byrne J, Gujar P, Kee K, Schüle B, Dolmetsch RE, Langston W, Palmer TD, Pera RR (2011) LRRK2 mutant iPSC-derived DA neurons demonstrate increased susceptibility to oxidative stress. Cell Stem Cell 8:267–280

Novitch BG, Chen AI, Jessell TM (2001) Coordinate regulation of motor neuron subtype identity and pan-neuronal properties by the bHLH repressor Olig2. Neuron 31:773–789

Pang ZP, Yang N, Vierbuchen T, Ostermeier A, Fuentes DR, Yang TQ, Citri A, Sebastiano V, Marro S, Sudhof TC, Wernig M (2011) Induction of human neuronal cells by defined transcription factors. Nature 476:220–223

Pfisterer U, Kirkeby A, Torper O, Wood J, Nelander J, Dufour A, Bjorklund A, Lindvall O, Jakobsson J, Parmar M (2011) Direct conversion of human fibroblasts to dopaminergic neurons. Proc Natl Acad Sci USA 108:10343–10348

Qiang L, Fujita R, Yamashita T, Angulo S, Rhinn H, Rhee D, Doege C, Chau L, Aubry L, Vanti WB, Moreno H, Abeliovich A (2011) Directed conversion of Alzheimer's disease patient skin fibroblasts into functional neurons. Cell 146:359–371

Roy NS, Cleren C, Singh SK, Yang L, Beal MF, Goldman SA (2006) Functional engraftment of human ES cell-derived dopaminergic neurons enriched by coculture with telomerase-immortalized midbrain astrocytes. Nat Med 12:1259–1268

Sekiya S, Suzuki A (2011) Direct conversion of mouse fibroblasts to hepatocyte-like cells by defined factors. Nature 475:390–393

Son EY, Ichida JK, Wainger BJ, Toma JS, Rafuse VF, Woolf CJ, Eggan K (2011) Conversion of mouse and human fibroblasts into functional spinal motor neurons. Cell Stem Cell 9:205–218

Szabo E, Rampalli S, Risueno RM, Schnerch A, Mitchell R, Fiebig-Comyn A, Levadoux-Martin M, Bhatia M (2010) Direct conversion of human fibroblasts to multilineage blood progenitors. Nature 468:521–526

Thier M, Worsdorfer P, Lakes YB, Gorris R, Herms S, Opitz T, Seiferling D, Quandel T, Hoffmann P, Nothen MM, Brüstle O, Edenhofer F (2012) Direct conversion of fibroblasts into stably expandable neural stem cells. Cell Stem Cell 10:473–479

Utikal J, Polo JM, Stadtfeld M, Maherali N, Kulalert W, Walsh RM, Khalil A, Rheinwald JG, Hochedlinger K (2009) Immortalization eliminates a roadblock during cellular reprogramming into iPS cells. Nature 460:1145–1148

Vierbuchen T, Ostermeier A, Pang ZP, Kokubu Y, Sudhof TC, Wernig M (2010) Direct conversion of fibroblasts to functional neurons by defined factors. Nature 463:1035–1041

Weintraub H, Tapscott SJ, Davis RL, Thayer MJ, Adam MA, Lassar AB, Miller AD (1989) Activation of muscle-specific genes in pigment, nerve, fat, liver, and fibroblast cell lines by forced expression of MyoD. Proc Natl Acad Sci USA 86:5434–5438

Wu H, Xu J, Pang ZP, Ge W, Kim KJ, Blanchi B, Chen C, Sudhof TC, Sun YE (2007) Integrative genomic and functional analyses reveal neuronal subtype differentiation bias in human embryonic stem cell lines. Proc Natl Acad Sci USA 104:13821–13826

Yamanaka S (2007) Strategies and new developments in the generation of patient-specific pluripotent stem cells. Cell Stem Cell 1:39–49

Yang D, Zhang ZJ, Oldenburg M, Ayala M, Zhang SC (2008) Human embryonic stem cell-derived dopaminergic neurons reverse functional deficit in Parkinsonian rats. Stem Cells 26:55–63

Yoo AS, Staahl BT, Chen L, Crabtree GR (2009) MicroRNA-mediated switching of chromatin-remodelling complexes in neural development. Nature 460:642–646

Yoo AS, Sun AX, Li L, Shcheglovitov A, Portmann T, Li Y, Lee-Messer C, Dolmetsch RE, Tsien RW, Crabtree GR (2011) MicroRNA-mediated conversion of human fibroblasts to neurons. Nature 476:228–231

Zhou Q, Melton DA (2008) Extreme makeover: converting one cell into another. Cell Stem Cell 3:382–388

Proposing a Model for Studying Primate Development Using Induced Pluripotent Stem Cells

Maria C.N. Marchetto, Alysson R. Muotri, and Fred H. Gage

Abstract New genomic tools provide us with high-resolution information about the alterations that may have resulted in the evolution of our own species. However, all information available to date for comparative studies between humans and our closest relatives, the non-human primates (NHP), comes from DNA/RNA samples extracted from preserved (post-mortem) tissues. These samples do not always fairly represent the distinctive traits of live cell development; nor do they represent cell behavior. Ideally, the identification of differences in genetic makeup between related species should be translated into phenotypical divergence. In this chapter, we will discuss the idea of developing and characterizing induced pluripotent stem cells (iPSC) from our closest relatives apes, such as bonobos, chimpanzees and gorillas. We then will discuss experimental protocols that will allow us to compare developing live neurons from humans to those from NHP and will suggest how to interpret possible outcomes in light of differences that have been previously involved in human speciation, such as brain size and differential gene expression. Such a culture model could provide new insights into human adaptation, with potential consequences for biomedical research and the basic biology of the species.

M.C.N. Marchetto • F.H. Gage
Laboratory of Genetics, The Salk Institute for Biological Studies, 10010 North Torrey Pines Road, La Jolla, CA 92037, USA
e-mail: gage@salk.edu

A.R. Muotri (✉)
Department of Pediatrics/Rady Children's Hospital San Diego, Department of Cellular & Molecular Medicine, Stem Cell Program, University of California San Diego, School of Medicine, 9500 Gilman Dr, La Jolla, CA 92093, USA
e-mail: muotri@ucsd.edu

F.H. Gage and Y. Christen (eds.), *Programmed Cells from Basic Neuroscience to Therapy*, Research and Perspectives in Neurosciences 20,
DOI 10.1007/978-3-642-36648-2_4, © Springer-Verlag Berlin Heidelberg 2013

Introduction

Anthropogeny, the science of explaining the origin of humans, utilizes different perspectives to provide insights into one of the greatest mysteries in evolution: the rise of the human lineage. Current views of human evolution, supported by fossil records, suggest the presence of many branches of the human lineage, but only one species survived (Wood and Collard 1999). The natural interest in understanding the process that makes us humans goes beyond mere anthropocentrism and philosophical debates about the human condition. Increased understanding of the differences between humans and our closest living ancestors, non-human primates (NHP) such as chimpanzees (*Pan troglodytes*), bonobos (*Pan paniscus*) and gorillas (*Gorilla gorilla*), is likely to contribute to biomedical advances. Important insights relevant to human health could be obtained. To mention a few examples, humans and NHP exhibit distinct progression rates of Acquired Immunodeficiency Syndrome (AIDS; Novembre et al. 1997), variable immunity against *Plasmodium falciparum* (Escalante and Ayala 1994), a prominent increase in neurofibrillary tangles typical of Alzheimer's Disease (Gearing et al. 1994) and variable cancer susceptibility (Seibold and Wolf 1973). A more comprehensive list can be found at the web-based "Museum of Comparative Anthropogeny," http://carta.anthropogeny.org/moca and in the publications cited therein (Varki 2000; Varki and Altheide 2005). Having different cell types from humans and NHP available for study could allow researchers to probe the nature of these differences in disease susceptibility between species and even shed light on new treatments.

Humans, chimpanzees and bonobos are genetically very similar, sharing nearly 99 % of their genomic sequences, yet it is not difficult to identify important features in which we are clearly distinct from them. Distinctions between humans and NHP become clearer at late stages of development. As the species develop, morphological differences become more and more evident. In fact, during the phylotypic stage, or early embryonic development, all primate embryos express a suite of characters that are substantially common, including a notochord, dorsal nerve cord, pharyngeal archers and blocks of somites. At this stage, pattern-forming genes, such as Hox genes, are first expressed, establishing the molecular blueprint of the body plan. Features that distinguish different primates groups, such as the cranium structure, only become apparent at later stages. Cognitive differences between humans, chimpanzees and bonobos become very distinct as the species develop. At a certain stage, human cognition becomes more pronounced when compared to NHP than the difference seen between chimpanzees and bonobos. Chimpanzees and bonobos develop very different cognitive abilities that are specifically related to their lifestyles and social organization. Remarkably, humans differ from other primates in the uniqueness of the size, interconnections and neuronal complexity of the brain, which allow for the development of sophisticated behaviors such as language, self-awareness, symbolic thought and cultural learning (Muotri and Gage 2006). The characterization of early neural developmental stages in primate brains is an important step toward understanding human brain development, disease and

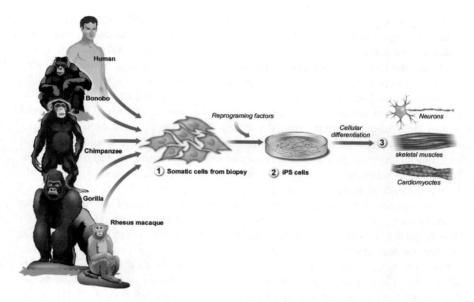

Fig. 1 Scheme of derivation and differentiation of induced pluripotent stem cells (iPSCs) from human and non-human-primate (NHP) species. The numbers indicate the steps involving in vitro modeling using reprogrammed cells from various species: (1) obtain somatic cells from biopsies (e.g., skin fibroblasts); (2) use reprogramming factors to derive new iPSC lines; (3) differentiate the iPSC in the target tissue (e.g., neuroprogenitors and neurons) and then perform comparative studies (refer to the text for examples of different assays)

evolution. We propose that a culture model using neurons derived from NHP induced pluripotent stem cells (iPSC) could provide new insights into human adaptation features and could have implications in biomedical research (Fig. 1).

Identifying Differences Between Human and NHP Brains

Accumulating evidence suggests that the evolution of the human brain, after the split from our common ancestors, was accompanied by discrete modifications in local circuitry and interconnectivity of selected parts of the brain (Schenker et al. 2005; Semendeferi et al. 2001). These modifications may be a consequence of selective changes that occurred in specific parts of the genome, affecting the phenotype of certain cell types. Modeling early stages of primate brain development would allow researchers to connect evolutionary genomic modifications to relevant physiological alterations in early embryogenesis.

Cell migration and brain organization. Cortical neurons in the developing mammalian brain migrate in a complex migratory process whereby each generation bypasses the previous one, in a phenomenon referred to as the "inside-out" gradient of neurogenesis (Rakic 2009). Although the biological significance of organized neuronal migration for cortex development is not clear, perturbations in its pattern often lead to abnormal cortical function (Gleeson and Walsh 2000). Accumulating evidence suggests that species-specific differences in cortical patterning and size originate in the early proliferative zones where neuroprogenitors reside (Cholfin and Rubenstein 2007; Lukaszewicz et al. 2006; O'Leary and Borngasser 2006). In theory, neuroprogenitor cell migration could be monitored using iPSC-derived neural cells from different primate species. We propose that detecting the species-specific differences in neural progenitor migration patterns could assist us in explaining/understanding the differences observed in brain size and organization. Identifying events that happen during brain development in different species could help to determine human-specific traits of neural organization.

Neuroanatomical studies on developing postmortem brains are beginning to identify those differences. The frontal polar part of the prefrontal cortex (PFC: also defined as Broadmann area 10 or BA10) is the largest cytoarchitectonic area in the human brain and it is believed to be involved in strategic processes in memory retrieval and executive function. During human evolution, increased function in this area resulted in its expansion relative to the rest of the brain. Area 10 in humans has the lowest neuron density among primate brains (Semendeferi et al. 2001). It has been proposed that, during hominid evolution, this area underwent a number of changes involving a considerable increase in overall size and a specific increase in connectivity, especially with other higher-order association areas (Semendeferi et al. 2001). Paleoanatomical observations of cranial endocasts taken from inside the skulls of human ancestral *Homo floresiensis* show an expansion in the frontal polar region, suggesting enlargement of its Brodmann's area 10 and indicating that this area was important during human evolution (Falk et al. 2005).

Neuroanatomical analysis of the developing brain could provide some insights into the processes that might have generated the differences found in adult brains. Analysis of dendritic patterns of human pyramidal neurons in the developing human cortex showed that the developmental time course of basilar dendritic systems was heterochronous and more protracted for BA10 than for areas BA4, BA3-2-1, and BA18 (Travis et al. 2005). Buxhoeveden et al. (2006) reported that a 2-year-old human possesses cortical minicolumns (vertical cortical columns that originate from neuroprogenitor cells) that are 90 % of the adult width in V1 (primary visual cortex) but less than 75 % of the adult width in the PFC. In contrast, a 2-year-old bonobo had PFC values within the range of the adult apes. It is thus reasonable to hypothesize that both the remarkable increase in human PFC values and their departure from the ape pattern took place sometime after the age of 2 years (Semendeferi et al. 2001). This increase is unique to the PFC and does not take place in any other area analyzed, where minicolumn widths remain at values similar to those seen in apes. Therefore, until at least the age of 2, humans share similar absolute minicolumn values with great apes in the PFC and other areas analyzed. Thus, it seems that a specific aspect of the neuronal phenotype (dendritic length and arborization) is susceptible to developmental and genetic alterations. All of the above findings suggest that alterations of the dendritic structure in a specific class of neurons (layer III pyramidal neurons in the PFC) may be involved in the uniquely human pattern of neuronal microcircuitry that underlies higher cognitive functions. Comparing these observations to data from developing neurons generated from primate iPSC cells could provide important insights into the developmental window in which the differences between humans and NHP brains occur and the underlying molecular bases for those differences.

Expression profile studies. The amplification and complexity of the cerebral neocortex during evolution are key features in human brain function. Cortical expansion accounts for a great deal of the difference in higher cognitive processing between humans and our closest living relatives. Despite the substantial genomic similarity, humans differ considerably from the other great apes in terms of brain function, cultural complexity and language acquisition. Because there is little evidence that simple addition or subtraction of genes is sufficient to explain such differences (The Chimpanzee and Analysis Consortium 2005; Hill and Walsh 2005), changes in the regulation (levels and patterns of expression) of genes shared between humans and chimpanzees have been proposed to play an important role in shaping neuronal networks and perhaps defining cognitive differences between the two species (Enard et al. 2002; King and Wilson 1975). Importantly, it has been suggested that human brain evolution is associated with changes in gene expression specifically within the brain as opposed to in other tissues such as liver (Enard et al. 2002). Nonetheless, most of the information available to date from comparative expression profile studies between humans and our closest living relatives comes from samples extracted from post-mortem tissues (Caceres et al. 2003; Enard et al. 2002; Marvanova et al. 2003; Oldham et al. 2006; Uddin et al. 2004).

iPSC technology could offer an additional tool for expression profile analysis during various steps of neuronal development/differentiation. Advanced whole genome sequencing technology facilitates detection of new mutations and the relevant variations in complex genetic diseases leading to common clinical outcomes. Combining in-depth DNA sequencing analysis with in vitro, species-specific neuronal phenotypes (such as cell migration, as mentioned above) could potentially provide a comprehensive database about neuronal features that could be used to tease out differences between species.

iPSC Technology and Neural Differentiation

Pluripotent cells from non-human primates. Pluripotent embryonic stem cells (ESC) have been successfully generated from the inner cell mass of blastocysts and can be induced to differentiate in vitro and in vivo into various cell types. Unfortunately, although ESC have being derived from humans (Thomson et al. 1998), rhesus monkeys and marmosets (Thomson et al. 1996; Thomson and Marshall 1998), limited accessibility to blastocysts from endangered species makes ESC derivation from chimpanzees, bonobos and gorillas a difficult task. Reprogramming of somatic cells to a pluripotent state by over-expression of specific genes (iPSC) has been accomplished using cells from several species, including mouse and human (Takahashi et al. 2007; Takahashi and Yamanaka 2006; Yu et al. 2007) and more recently from rhesus monkey (Liu et al. 2008), short-tailed monkey (drill; Ben-Nun et al. 2011), rat (Li et al. 2009; Liao et al. 2009) and pig (Ezashi et al. 2009; Wu et al. 2009). The resultant iPSC carry a similar genetic background as the donor individual. Isogenic pluripotent cells are attractive to medicine not only for their potential therapeutic purpose – with lower risk of immune rejection – but also for understanding complex diseases with heritable and sporadic conditions (Muotri 2009).

Another proposed use for iPSC derived from NHP is to preserve the genetic material of endangered species. The drill is one of Africa's most endangered mammals. In the wild, drills are found only in small areas of Nigeria, Southwestern Cameroon and Equatorial Guinea. The number of drills is declining as a result of illegal bush meat commerce and habitat destruction. Ben-Nun et al. (2011) derived iPSC from the African primate drill (*Mandrillus leucophaeus*) and those cells were karyotypically normal and were able to self renew and differentiate into the three embryonic layers in culture. The authors propose that preserving the genomes of endangered species in the form of iPSC would be useful for generating iPSC-derived germ cells to assist reproduction efforts. Even though substantial challenges still remain, success in generating and differentiating iPSC from endangered species of NHP could contribute to conservation efforts in addition to assisting evolutionary studies.

Protocols for neural lineage speciation. New neuronal differentiation protocols that can obtain particular subtypes of neurons from iPSC are already available (e.g., dopaminergic, hippocampal and cholinergic neurons) and are currently being used for disease modeling purposes (Kriks et al. 2011). It remains to be determined whether the neurons that can be developed in culture are the ones relevant for the differences observed in brain function and anatomy between species. Nonetheless, improving the protocols for generating more homogeneous cultures and more functionally mature neurons would be extremely useful to detect subtle, but relevant, differences. More comparative neuroanatomical data on postmortem brain tissues will be needed to help define the specific neuronal subtypes and brain regions that have the most relevant differences between species.

Phenotypical Assays for Comparative Studies

Improving the protocols for evaluating neuronal connectivity properties, synaptic plasticity and electrophysiological functional outcomes will definitely be of great value in detecting species-specific phenotypes. Examples of techniques that are already available are calcium imaging, light-activated channel rhodopsins, uncaged glutamate, transsynaptic labeling using virus, multielectrode arrays, high-resolution live imaging for spine motility and maturation, synaptic protein recruitment and axonal transporting dynamic visualization. Combining ideas and technologies from established fields will be highly beneficial to this nascent field. A practical example is the recent incorporation of biomaterials and bioengineering techniques for improved differentiation of iPSC cultures. New alternative methods for better compartmentalization and isolation of neuronal processes using micro fluidic chambers have been explored and implemented in primary neural cultures. Compartmentalization of neurons using engineered devices would allow comparisons of the dynamic behavior and molecular anatomy of neurons in culture (Taylor et al. 2005; Wissner-Gross et al. 2011). Additionally, using engineered tridimensional bio-matrices to simulate tissue structures may more authentically recapitulate in vivo neuronal branching and connectivity and may potentiate a more complete in vitro maturation.

A useful alternative for obtaining more mature and integrated neurons is in vivo grafting of neural progenitor cells in rodent brains (Jensen et al. 2011; Muotri et al. 2005). Studying the anatomy and function of transplanted neurons over time informs studies of the neurodevelopmental aspects of a disease as well as of cell-autonomous versus non-autonomous elements. Developmental hallmarks such as neuronal pruning, dendritic branching, spine formation and maturation could be dynamically observed as transplanted neural progenitors differentiate into neurons over time. State-of-the-art intracranial live imaging techniques coupled with electrophysiological studies could facilitate studies of functional integration properties from neurons in real time. Transplantation may also lead to the generation of specific subtypes of neurons that are difficult to produce through in vitro differentiation protocols. These transplantation studies could ideally recapitulate the in vivo characteristics of cells more authentically.

Conclusion and Perspectives

Here we propose the use of a novel evolutionary tool, iPSC technology, to provide insights into human evolution. To identify potential differences during early stages of development and to highlight human-specific mechanisms, we propose to generate iPSC from NHP (e.g., chimpanzees, bonobos, gorillas and rhesus) skin fibroblasts. We believe that the iPSC system described here will allow for the investigation of cellular and molecular mechanisms present in several tissues that distinguish the species studied. We have focused on the nervous system and propose that iPSC could undergo early stages of neural development in culture, generating different subtypes of neurons that are involved in different brain functions that are relevant for human speciation. Resultant iPSC could recapitulate early stages of neural development of genetically close species of primates with very distinct morphological, behavioral and cognitive features. The dissection of the cellular and molecular mechanisms that distinguish humans from our closest relatives at early stages of embryogenesis and in specific type of cells (e.g., neurons) is likely to become a new resource for evolutionary studies, with practical implications for human health. Finally, another positive effect of this research is that, as molecular and cellular links between humans and "great apes" are discovered, our society may become more motivated in its conservation efforts.

References

Ben-Nun IF, Montague SC, Houck ML, Tran HT, Garitaonandia I, Leonardo TR, Wang YC, Charter SJ, Laurent LC, Ryder OA, Loring JF (2011) Induced pluripotent stem cells from highly endangered species. Nat Methods 8:829–831

Buxhoeveden DP, Semendeferi K, Buckwalter J, Schenker N, Switzer R, Courchesne E (2006) Reduced minicolumns in the frontal cortex of patients with autism. Neuropathol Appl Neurobiol 32:483–491

Caceres M, Lachuer J, Zapala MA, Redmond JC, Kudo L, Geschwind DH, Lockhart DJ, Preuss TM, Barlow C (2003) Elevated gene expression levels distinguish human from non-human primate brains. Proc Natl Acad Sci USA 100:13030–13035

Cholfin JA, Rubenstein JL (2007) Patterning of frontal cortex subdivisions by Fgf17. Proc Natl Acad Sci USA 104:7652–7657

Enard W, Khaitovich P, Klose J, Zollner S, Heissig F, Giavalisco P, Nieselt-Struwe K, Muchmore E, Varki A, Ravid R, Doxiadis GM, Bontrop RE, Pääbo S (2002) Intra- and interspecific variation in primate gene expression patterns. Science 296:340–343

Escalante AA, Ayala FJ (1994) Phylogeny of the malarial genus Plasmodium, derived from rRNA gene sequences. Proc Natl Acad Sci USA 91:11373–11377

Ezashi T, Telugu BP, Alexenko AP, Sachdev S, Sinha S, Roberts RM (2009) Derivation of induced pluripotent stem cells from pig somatic cells. Proc Natl Acad Sci USA 106:10993–10998

Falk D, Hildebolt C, Smith K, Morwood MJ, Sutikna T, Brown P, Jatmiko, Saptomo EW, Brunsden B, Prior F (2005) The brain of LB1, *Homo floresiensis*. Science 308:242–245

Gearing M, Rebeck GW, Hyman BT, Tigges J, Mirra SS (1994) Neuropathology and apolipoprotein E profile of aged chimpanzees: implications for Alzheimer disease. Proc Natl Acad Sci USA 91:9382–9386

Gleeson JG, Walsh CA (2000) Neuronal migration disorders: from genetic diseases to developmental mechanisms. Trends Neurosci 23:352–359

Hill RS, Walsh CA (2005) Molecular insights into human brain evolution. Nature 437:64–67

Jensen MB, Yan H, Krishnaney-Davison R, Al Sawaf A, Zhang SC (2011) Survival and differentiation of transplanted neural stem cells derived from human induced pluripotent stem cells in a rat stroke model. J Stroke Cerebrovasc Dis 10. Doi: org/10.1016/j.jstrokecerebrovasdis.2011.09.008 (Pub ahead of print)

King MC, Wilson AC (1975) Evolution at two levels in humans and chimpanzees. Science 188:107–116

Kriks S, Shim JW, Piao J, Ganat YM, Wakeman DR, Xie Z, Carrillo-Reid L, Auyeung G, Antonacci C, Buch A, Yang L, Beal MF, Surmeier DJ, Kordower JH, Tabar V, Studer L (2011) Dopamine neurons derived from human ES cells efficiently engraft in animal models of Parkinson's disease. Nature 480:547–551

Li W, Wei W, Zhu S, Zhu J, Shi Y, Lin T, Hao E, Hayek A, Deng H, Ding S (2009) Generation of rat and human induced pluripotent stem cells by combining genetic reprogramming and chemical inhibitors. Cell Stem Cell 4:16–19

Liao J, Cui C, Chen S, Ren J, Chen J, Gao Y, Li H, Jia N, Cheng L, Xiao H, Xiao L (2009) Generation of induced pluripotent stem cell lines from adult rat cells. Cell Stem Cell 4:11–15

Liu H, Zhu F, Yong J, Zhang P, Hou P, Li H, Jiang W, Cai J, Liu M, Cui K, Qu X, Xiang T, Lu D, Chi X, Gao G, Ji W, Ding M, Deng H (2008) Generation of induced pluripotent stem cells from adult rhesus monkey fibroblasts. Cell Stem Cell 3:587–590

Lukaszewicz A, Cortay V, Giroud P, Berland M, Smart I, Kennedy H, Dehay C (2006) The concerted modulation of proliferation and migration contributes to the specification of the cytoarchitecture and dimensions of cortical areas. Cereb Cortex 16(Suppl 1):i26–i34

Marvanova M, Menager J, Bezard E, Bontrop RE, Pradier L, Wong G (2003) Microarray analysis of nonhuman primates: validation of experimental models in neurological disorders. FASEB J 17:929–931

Muotri AR (2009) Modeling epilepsy with pluripotent human cells. Epilepsy Behav 14(Suppl 1):81–85

Muotri AR, Gage FH (2006) Generation of neuronal variability and complexity. Nature 441:1087–1093

Muotri AR, Nakashima K, Toni N, Sandler VM, Gage FH (2005) Development of functional human embryonic stem cell-derived neurons in mouse brain. Proc Natl Acad Sci USA 102:18644–18648

Novembre FJ, Saucier M, Anderson DC, Klumpp SA, O'Neil SP, Brown CR 2nd, Hart CE, Guenthner PC, Swenson RB, McClure HM (1997) Development of AIDS in a chimpanzee infected with human immunodeficiency virus type 1. J Virol 71:4086–4091

O'Leary DD, Borngasser D (2006) Cortical ventricular zone progenitors and their progeny maintain spatial relationships and radial patterning during preplate development indicating an early protomap. Cereb Cortex 16(Suppl 1):i46–i56

Oldham MC, Horvath S, Geschwind DH (2006) Conservation and evolution of gene coexpression networks in human and chimpanzee brains. Proc Natl Acad Sci USA 103:17973–17978

Rakic P (2009) Evolution of the neocortex: a perspective from developmental biology. Nat Rev Neurosci 10:724–735

Schenker NM, Desgouttes AM, Semendeferi K (2005) Neural connectivity and cortical substrates of cognition in hominoids. J Human Evol 49:547–569

Seibold HR, Wolf RH (1973) Neoplasms and proliferative lesions in 1065 nonhuman primate necropsies. Lab Anim Sci 23:533–539

Semendeferi K, Armstrong E, Schleicher A, Zilles K, Van Hoesen GW (2001) Prefrontal cortex in humans and apes: a comparative study of area 10. Am J Phys Anthropol 114:224–241

Takahashi K, Yamanaka S (2006) Induction of pluripotent stem cells from mouse embryonic and adult fibroblast cultures by defined factors. Cell 126:663–676

Takahashi K, Tanabe K, Ohnuki M, Narita M, Ichisaka T, Tomoda K, Yamanaka S (2007) Induction of pluripotent stem cells from adult human fibroblasts by defined factors. Cell 131:861–872

Taylor AM, Blurton-Jones M, Rhee SW, Cribbs DH, Cotman CW, Jeon NL (2005) A microfluidic culture platform for CNS axonal injury, regeneration and transport. Nat Meth 2:599–605

The Chimp Sequencing and Analysis Consortium (2005) Initial sequence of the chimpanzee genome and comparison with the human genome. Nature 437:69–87

Thomson JA, Marshall VS (1998) Primate embryonic stem cells. Curr Top Dev Biol 38:133–165

Thomson JA, Kalishman J, Golos TG, Durning M, Harris CP, Hearn JP (1996) Pluripotent cell lines derived from common marmoset (*Callithrix jacchus*) blastocysts. Biol Reprod 55:254–259

Thomson JA, Itskovitz-Eldor J, Shapiro SS, Waknitz MA, Swiergiel JJ, Marshall VS, Jones JM (1998) Embryonic stem cell lines derived from human blastocysts. Science 282:1145–1147

Travis K, Ford K, Jacobs B (2005) Regional dendritic variation in neonatal human cortex: a quantitative Golgi study. Dev Neurosci 27:277–287

Uddin M, Wildman DE, Liu G, Xu W, Johnson RM, Hof PR, Kapatos G, Grossman LI, Goodman M (2004) Sister grouping of chimpanzees and humans as revealed by genome-wide phylogenetic analysis of brain gene expression profiles. Proc Natl Acad Sci USA 101:2957–2962

Varki A (2000) A chimpanzee genome project is a biomedical imperative. Genome Res 10:1065–1070

Varki A, Altheide TK (2005) Comparing the human and chimpanzee genomes: searching for needles in a haystack. Genome Res 15:1746–1758

Wissner-Gross ZD, Scott MA, Ku D, Ramaswamy P, Fatih Yanik M (2011) Large-scale analysis of neurite growth dynamics on micropatterned substrates. Integr Biol (Camb) 3:65–74

Wood B, Collard M (1999) The human genus. Science 284:65–71

Wu Z, Chen J, Ren J, Bao L, Liao J, Cui C, Rao L, Li H, Gu Y, Dai H, Zhu H, Teng X, Cheng L, Xiao L (2009) Generation of pig induced pluripotent stem cells with a drug-inducible system. J Mol Cell Biol 1:46–54

Yu J, Vodyanik MA, Smuga-Otto K, Antosiewicz-Bourget J, Frane JL, Tian S, Nie J, Jonsdottir GA, Ruotti V, Stewart R, Slukvin II, Thomson JA (2007) Induced pluripotent stem cell lines derived from human somatic cells. Science 318:1917–1920

HTT Evolution and Brain Development

Chiara Zuccato and Elena Cattaneo

Abstract Huntingtin (htt) is the 800 million-year-old 3,144 amino acid protein product of the Huntington's disease (HD) gene, which carries a tri-nucleotide CAG repeat then translated into polyglutamine (polyQ) at an evolutionarily conserved NH_2-terminal position in exon 1. The CAG triplet is polymorphic in the normal population, ranging from 9 to 32 repetitions. In humans, an expansion of the repeats to more than 35 causes HD, a fatal, genetically dominant neurodegenerative disorder (MacDonald et al., Cell 72:971–983, 1993).

Since the discovery of the HD gene in 1993, knowledge of normal htt function has been fragmented. In mammals, htt is expressed in the early post-fertilization stages and becomes enriched in the developing and adult brain, where it promotes the transcription of neuronal genes, vesicle trafficking and axonal transport. Htt also acts as an anti-apoptotic protein in vivo in the brain and in cultured neural and peripheral cells. Subversion or exacerbation of these functions by an abnormally expanded polyQ repeat contributes to neuronal vulnerability in HD and suggests that loss of normal htt function may be implicated in HD (Cattaneo et al. 2005; Zuccato et al. 2010).

During embryogenesis, htt is critical for gastrulation and neurogenesis. When htt expression is experimentally reduced to below 50 % of wild-type levels, defects in the epiblast, the structure that gives rise to the neural tube, are observed (White et al. 1997). In the htt knock-down zebrafish embryo, defects are found in the most anterior regions of the neural plate (Henshall et al. 2009). Later in development, neuroblasts in the telencephalon must synthesize htt to progress correctly through differentiation (Reiner et al. 2001). This process might depend on a recent finding that htt regulates mitotic spindle orientation in the developing mammalian cortex,

C. Zuccato • E. Cattaneo (✉)
Centre for Stem Cell Research, Università degli Studi di Milano, Via Balzaretti 9,
Milan 20133, Italy
e-mail: elena.cattaneo@unimi.it

F.H. Gage and Y. Christen (eds.), *Programmed Cells from Basic Neuroscience to Therapy*, Research and Perspectives in Neurosciences 20,
DOI 10.1007/978-3-642-36648-2_5, © Springer-Verlag Berlin Heidelberg 2013

an activity that can affect cortical progenitor cell fate decisions (Godin et al. 2010). Despite this knowledge, the exact cellular and molecular functions that made htt indispensable for neural tube formation and brain morphogenesis remained largely obscure.

To study htt function during early brain development, we recently used an approach that combines neurodevelopmental and evolutionary studies. Our final goal was to infer information about functional regions of htt that have emerged during evolution and have become important for the development of the nervous system.

In this chapter, we discuss the discoveries that have mapped the evolutionary history of htt and a recent work from our group that highlights a critical role for the protein in brain evolution and development. Htt originated 800 million years ago, before the protostome-deuterostome divergence. During deuterostome evolution, htt acquired in chordates a unique activity that was critical for neurulation. This function became progressively more specialized in organisms that were characterized by a more evolved, centralized nervous system. Accordingly, the polyQ tract in htt, which is peculiar to deuterostomes, may contribute to the evolutionary transition from invetebrates to vertebrates and mark the acquisition of centralized nervous systems in primitive chordates. These findings suggest that htt (and its polyQ) may be a developmental factor that has acted during evolution by regulating aspects of nervous system formation, with neurulation being one of the first decisive events. Finally, we describe how this evolutionary study has revealed a novel target for htt during neurulation, with implications also for the adult HD brain.

The Evolutionary History of htt

The origin of the *hd* gene dates back to 800 million years before the protostome-deuterostome divergence (Palidwor et al. 2009). In 2009, the group of M. Andrade at the Max-Delbruck Center for Molecular Medicine in Berlin reported that htt was present in the ancient organism *Dictyostelium discoideum* (slime mold) and that no homologs were found in fungi (*Saccharomices cerevisiae*) or in plants (Palidwor et al. 2009).

The *hd* locus in *Dictyostelium* encodes for a protein composed of 3,095 amino acids, comparable to the 3,144 amino acids of human htt although no Qs were found in its NH_2-terminal portion (Palidwor et al. 2009; Myre et al. 2011). Neverthless, a polyQ tract of 19 residues, which is comparable in size to that found in normal-range human htt, is present in the *Dictyostelium* gene, but it is located further downstream of the initiator methionine (at residue 533 rather than at residue 18, as in the human protein). This finding could be expected because of the high number of predicted proteins (~34 %) that contain homopolymer tracts of 15 residues or more in *Dictyostelium* (Myre et al. 2011; Palidwor et al. 2009). These studies suggested that the polyQ in htt NH$_2$-terminus was acquired later in evolution.

Dictyostelium is a social amoeba that lives in forest soil where it hunts bacteria and yeast. When hunting is not successful, this unicellular organism becomes one multicellular entity and forms a fruiting body to disperse spores. Genes can be knocked out rapidly by targeted homologous recombination in *Dictyostelium*. Myre and colleagues (2011) at Massachusetts General Hospital found that inactivation of the htt gene in *Dictyostelium* was compatible with cell growth but produced cell-autonomous phenotypes. Indeed the mutant cells could not complete those coordinated and synchronous events that lead to the development of a multicellular organism. Loss of htt in *Dictyostelium* impacted upon many biochemical processes, including chemotaxis, cytokinesis, cell shape, and homophilic cell–cell adhesion during hypo-osmotic stress (Wang et al. 2011; Myre et al. 2011).

Recent work from our laboratory showed that htt from *Dictyostelium* is anti-apoptotic, as in mammals. Previous studies have found that mammalian htt has an anti-apoptotic effect in neural and non-neural cells in vitro and in the brain (Zuccato et al. 2010) and that this activity is embedded in the htt 548 amino acid NH$_2$-terminus (Rigamonti et al. 2000). These findings suggest that the ability to regulate apoptosis may be one of htt's ancestral activities (Zuccato et al. 2010; Lo Sardo et al. 2012).

All of this evidence led to the proposition that htt appeared in a common ancestor of the protostomes and deuterostomes to specifically exert non-neuronal functions, which can be expected because a primitive organism such as *Dictyostelium* has no neurons and, obviously, no brain.

Htt in Protostomes

Our knowledge of Htt function(s) in protostomes comes from studies in flies and, particularly, in the *Drosophila* lineage. *Drosophila melanogaster* htt (*dhtt*) was identified in 1999 in work by Li and colleagues (1999) at Stanford University, who found that the predicted htt protein had 3,583 amino acids and was several hundred amino acids larger than any other previously characterized htt protein. Analysis of the genomic and cDNA sequences indicated that *dhtt* has 29 exons, compared with the 67 exons present in vertebrate htt genes, and that it lacks the polyQ and polyP stretches present in its mammalian counterparts. Li et al. (1999) also reported that *dhtt* was expressed ubiquitously during all stages of the fly development.

Our bioinfomatic analyses of htt from *Drosophila melanogaster, Drosophila pseudoobscura, Apis mellifera* and *Tribolium castaneum* revealed that the htt amino acid sequence in insects followed a more heterogeneous evolution compared to deuterostomes (Tartari et al. 2008). Andrade's group recently noted that, in various *Drosophilae,* htt has several polyQ stretches in other regions of the protein (e.g., *Drosophila yakuba* has 10 glutamines at positions 625–634 and 12 glutamines in a stretch of 14 amino acids at positions 1118–1131), which are absent in the human protein. This evidence points towards the possibility that htt in *Drosophilae* lineages experienced independent events of insertion of polyQ tracts (Schaefer et al. 2012). The function of these polyQ tracts is currently unknown.

To explore htt's function in *Drosophila,* Zhang and colleagues (2009) from Harvard Medical School in Boston generated a null mutant in the putative *dhtt* homolog and found that the larva develops without any defect in the gastrulation process. In contrast to what happens in mammals, htt in *Drosophila* is therefore not involved in controlling embryogenesis (Zhang et al. 2009; Zuccato et al. 2010). This phenotypic discrepancy might reflect intrinsic differences in mouse and fly embryonic development (Li et al. 1999; Tartari et al. 2008). It is also possible that a redundant pathway might compensate for the loss of *dhtt*. As suggested by Zheng and Joinnides (2009) from the University of Pennsylvania, there may also be subtle defects in *dhtt-ko* flies that have escaped detection under normal conditions and that might intensify with stress (e.g., aging, environmental insults, oxidative stress, injury). Further analyses revealed a mild phenotype in the fly adult brain. *Dhtt* knock-out mutants showed reduced numbers of branches in the axonal termini of giant fiber neurons, but they did not show any defects in synapse formation, neurotransmission or axonal transport (Zhang et al. 2009). This finding seems to disagree with the previously reported ability of *dhtt* to control axonal transport and synapse formation (Gunawardena et al. 2003). Zhang and colleagues (2009) also revealed that *dhtt* mutants exhibited significantly reduced mobility and viability as they aged, suggesting a role for the protein in maintaining the long-term functioning and survival of adult flies (Zhang et al. 2009).

Recent findings have revealed that two important functions of htt in mammals are conserved in flies. A study from the group of S. Humbert at Institut Curie in Orsay showed that *dhtt* controls mitotic spindle orientation and cortical neurogenesis in mice (Godin et al. 2010). *Dhtt* could functionally replace the missing mammalian htt protein, emphasizing the evolutionary conservation of this activity of the protein (Godin et al. 2010).

Recent work from our laboratory has shown that *dhtt* is anti-apoptotic in a cell death assay in htt-depleted mammalian embryonic stem (ES) cells (Lo Sardo et al. 2012). As previously mentioned, the anti-apoptotic activity of htt is evolutionarily old, being present at the base of the protostome–deuterostome divergence in the common ancestor *Dictyostelium* (Lo Sardo et al. 2012). This function has been maintained in flies, suggesting its conservation during evolution.

More studies are needed to understand the function of htt in flies. A genetic modifiers approach could be a valid option to identify genes and pathways that interact with *dhtt*. Investigation of htt functions in other protostomes such as flatworms, annelids and molluscs, in which the presence of the gene has been predicted, could represent an important step forward towards understanding how the *hd* gene has evolved functionally in that branch.

Htt in Deuterostomes

Bioinformatic analyses and the cloning of htt from old organisms that represent key points of deuterostome evolution have revealed important clues about the protein sequence, its domains and possible activities. In 2003, when our study on the evolution of htt began, most of the known htt protein homologs in deuterostomes belonged to vertebrates. At that time, *Homo sapiens* and the *Fugu* fish represented the most divergent vertebrates from which htt had been cloned (Sathasivam et al. 1997), but htt protein showed a high degree of conservation in these species, thus offering limited insights into the understanding of the protein domains and of the critical amino acid sequences emerging during evolution.

An important step in the reconstruction of htt's function along deuterostome evolution came with the cloning of htt from the echinoderm *Strongylocentrotus purpuratus* (Tartari et al. 2008), urochordate *Ciona intestinalis* (Gissi et al. 2006) and cephalocordate *Branchistoma floridae* (Candiani et al. 2007), which are key points of the phylogenetic tree for the development and evolution of the nervous system. When the htt sequence of these old deuterostomes was introduced into htt protein multialignment studies, which included huntingtin from vertebrates and insects, three putative htt domains could be identified and the history of the polyQ started to emerge. Moreover, the evidence that the Q length had progressively increased during deuterostome evolution, coinciding with the appearance and structuring of progressively more complex nervous systems, has led us to consider the possibility that the two events may be somehow linked.

Origin and Evolution of the htt polyQ in Deuterostomes

The appearance of the Q in htt dates back to the sea urchin (*Strongylocentrotus purpuratus*), which carries an NHQQ sequence at the NH_2 terminus of the protein (Tartari et al. 2008). This discovery predates an ancestral polyQ sequence in a non-chordate environment and defines the polyQ characteristic as being typical of the deuterostome branch. A study we conducted in collaboration with M. Pestarino and S. Candiani from the University of Genova demonstrated that two glutamines (QQ) are also present in htt from cephalochordate *Branchiostoma floridae* (Candiani et al. 2007). Nevertheless, one difference between sea urchin and amphioxus is that, in the latter, the 2Q are preceded by the hydrophobic amino acids AF, whereas in sea urchin the hydrophilic amino acids NH are present instead. Moreover, the first NH_2-terminal 15 amino acids of amphioxus are identical to those of vertebrates, whereas the same portion of the protein in sea urchin exhibits several differences in amino acid composition (Fig. 1). This observation suggests that the NH_2-terminal htt and its polyQ might have been subjected to a selective evolutionary pressure in primitive chordates that might have resulted in stabilization of the polyQ trait in amphioxus (Candiani et al. 2007; Tartari et al. 2008). Interestingly, htt of *Ciona intestinalis* genus lost the polyQ tract (Gissi et al. 2006). In this species, a single H

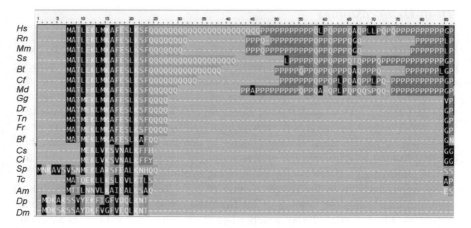

Fig. 1 The evolution of the polyQ region in huntingtin. Details of the multiple alignment (NH$_2$-terminus amino acid sequences) are listed following the phylogenetic tree. Species abbreviations are the following: *Homo sapiens* (*Hs*); *Rattus norvegicus* (*Rn*); *Mus musculus* (*Mm*); *Sus scrofa* (*Ss*); *Bos taurus* (*Bt*); *Canis familiaris* (*Cf*); *Monodelphis domestica* (*Md*); *Gallus gallus* (*Gg*); *Danio rerio* (*Dr*); *Tetraodon nigroviridis* (*Tn*); *Fugu rubripes* (*Fr*); *Branchiostoma floridae* (*Bf*); *Ciona savignyi* (*Cs*); *Ciona intestinalis* (*Ci*); *Strongylocentrotus purpuratus* (*Sp*); *Tribolium castaneum* (*Tc*); *Apis mellifera* (*Am*); *Drosophila pseudoobscura* (*Dp*); and *Drosophila melanogaster* (*Dm*)

or T residue is located at the position corresponding to the vertebrate polyQ stretch, suggesting low selective constraints acting on this region in the ascidian protein (Gissi et al. 2006; Fig. 1). The htt protein in *Ciona* is notably shorter than their vertebrate homologs, and this is probably due to deletions in the NH$_2$-terminal region of the protein (Gissi et al. 2006). Although *Ciona* possesses the main chordate traits during neural development, these differences can be expected since they quickly diverged from the other chordates and their genome is not representative of the ancestral chordate genome, which may reflect adaptation of the specific ecological niche of urochordates (Hughes and Friedman 2005).

The polyQ then expands in vertebrates. Four glutamines are present in fishes, amphibians, and birds. Six glutamines are found in htt from marsupial *Monodelphis domestica* and *Opossum*. The polyQ increases in size gradually in mammals. Seven glutamines are found in mouse, 8 in rat, 10 in dog, 15 in *Bos taurus* and 18 in pig. The analyses of the Q trait in primates from 10 different species revealed that the numbers of glutamines were remarkably consistent between species, with the combined range spanning from 7 to 16 Q (Rubinsztein et al. 1994). The human CAG repeats in the gene encoding for htt may have thus originated from a shorter ancestral (sequence) because of mutational bias, which causes expansion towards longer alleles (Rubinsztein et al. 1994). In humans, the polyQ tract reaches the longest and most polymorphic length (from 9 to 32; Fig. 1).

These observations indicate that the appearance of the polyQ trait in htt is evolutionarily important. Moreover, they point to the possibility that the emergence of the polyQ and its increase in length during deuterostome evolution may have coincided with the appearance of a more complex nervous system along this branch. Accordingly, the polyQ and/or flanking amino acids may influence brain development and evolution. CAG repeats are mutation-prone DNA tracts that are well-tolerated, or even encouraged, during evolution (Arthur 2004; Koren and Trifonov 2011; Kashi et al. 1997; Ruden et al. 2005). Recent studies have established that the specific "mutational" mechanism that led to a progressive increase in CAG tracts during evolution was favored by selection, possibly because this polymorphism contributed to the generation of widespread quantitative variation in important phenotypic traits. This theory assumes that the normal variation in the length of the CAG tract in genes involved in development may be implicated in morphological differences between species and phenotypic variations that influence normal brain structure (Fondon et al. 2008). Interestingly, recent data indicate that normal subjects carrying longer CAG repeats in the normal range have increased gray matter (Muhlau et al. 2012). Thus, glutamine repeats may introduce a bias into htt evolution, affecting embryo development and leading deuterostomes to acquire a nervous system of increased complexity (Lo Sardo et al. 2012).

But what is the functional significance of an expanded polyQ tract in htt during evolution? Earlier indications demonstrated that the polyQ mediates htt interaction with other proteins, as Nobel laureate Max Perutz discovered in 1994 (Perutz et al. 1994). Recent analyses from Andrade and colleagues confirmed these findings and showed that polyQ regions stabilize protein–protein interactions (Schaefer et al. 2012). PolyQ proteins are often involved in transcriptional regulatory complexes and protein-gene regulatory interactions that influence phenotypic variability (Whan et al. 2010). Therfore, it is possible that the polyQ may regulate htt function during evolution by interacting with different transcription factors and/or gene pathways that are critical for brain development (Zuccato et al. 2010). Very little information is available from mice in which the polyQ has been deleted in the mouse *Hdh* gene (Clabough and Zeitlin 2006). These mice exhibit significative defects in learning and memory tests, although development appears to be normal (Clabough and Zeitlin 2006).

Htt Domains: A Focus on the NH$_2$-Terminus

The discovery of a large panel of interactors has led to the hypothesis that htt might have a flexible structure capable of assuming specific conformations and activities during development and in adulthood (MacDonald 2003). Htt contains HEAT repeats that are implicated in protein–protein interactions (Andrade and Bork 1995), thus supporting the notion that the protein exerts its functions through different protein partners. Moreover, findings from our group have indicated that the first NH$_2$-terminal 548 amino acids of human htt are endowed with anti-apoptotic and neuroprotective activities, further suggesting that htt may use protein domains to exert its different activities (Rigamonti et al. 2000; Zuccato et al. 2003; Lo Sardo et al. 2012).

Our bioinformatic studies predicted the presence of three putative domains in htt, which consists of three major conserved regions corresponding to blocks 1–386 (htt1), 683–1,586 (htt2), and 2,437–3,078 (htt3) of human htt (Tartari et al. 2008). A comparison of more divergent htt orthologs and quantification of the evolutionary pressure on the three blocks revealed that the htt NH_2-terminal fragment (htt1) is the most recently evolved portion of the protein. This is the portion of the protein that bears the polyQ tract. We observed also that, when the polyQ length increases and couples with the adjacent polyP tract in mammals, conservation of htt1 becomes more stringent, thus further qualifying it as a putative protein domain (Tartari et al. 2008). It has been suggested that polyP stabilizes the polyQ tract by keeping it soluble, and it is interesting to note that this tract has emerged in parallel with longer polyQ stretches (Steffan et al. 2004; Ramazzotti et al. 2012).

Based on this knowledge, we have tested the hypothesis that the polyQ tract has impressed into the htt NH_2-terminus specific functions related to brain development during deuterostome evolution. By coupling neurodevelopmental and evolutionarily approaches, we found that htt is implicated in neurulation. Notably, we found that this function typically emerges in cephalochordates.

The Emerging Pro-neurulation Function of htt During Deuterostome Evolution

The first indication of a direct role for htt in neurulation came from experiments conducted in htt-depleted mouse ES cells. Wild-type ES cells subjected to a specific neural differentiation protocol generate tridimensional structures called "neuroepithelial rosettes" (Ying et al. 2003). These rosettes appear after 8 days of in vitro differentiation and are thought to recapitulate in vitro several aspects of neural tube development and early neurulation events (Elkabetz et al. 2008; Abranches et al. 2009). We found that loss of htt impairs this process, thus leading to cultures that were nearly devoid of properly formed neuroepithelial rosettes (Lo Sardo et al. 2012). This phenotype was termed *rosetteless* (Fig. 2). Analysis in the htt knock-down zebrafish embryos confirmed the role of htt in neurulation. Embryos injected with htt morpholino (MO) displayed an altered structure of the neural tube at the level of the diencephalon, with clusters of mis-positioned cells and cellular aggregates in the brain ventricles. Loss of htt led also to defects in apico-basal polarity during neurulation, which compromised the integrity of the neural tube (Lo Sardo et al. 2012).

These experiments also demonstrated that the domain of htt that is involved in driving neurulation is contained within the first ~500 amino acids of the protein. We found that NH_2-terminus (htt1) but not the C-terminus (htt3) of the *Mus musculus* htt produced a full rescue of the rosetteless phenotype in functional complementation assays performed in htt-depleted ES cells (Lo Sardo et al. 2012).

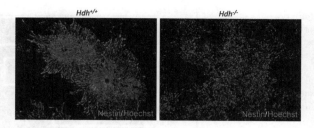

Fig. 2 Wild-type ($Hdh^{+/+}$) and htt-null ES cells ($Hdh^{-/-}$) were subjected to a monolayer neural differentiation protocol that converts embryonic stem (ES) cells into neuroepithelial progenitors that resemble those present at the time of neural plate closure and neural tube formation in vivo (Ying et al. 2003). Wild-type cells formed neuroepithelial rosettes whereas Hdh−/− cells generated neuroepithelial precursors with aberrant spatial organization and cultures devoid of properly formed rosettes. Cells were immunostained with Nestin/Hoechst

We also found that htt pro-neurulation activity is a recent acquisition in deuterostome evolution. The NH_2-terminal htt fragments from the evolutionarily distant *Dictyostelium* were not effective in rescuing the rosetteless phenotype. This finding further supports the notion that *Dictyostelium* htt carries non-neuronal functions (Myre et al. 2011). The ability of NH_2-terminal htt fragments to sustain neurulation was subtle in lower deuterostomes sea urchin and *Ciona* and then surprisingly appeared in the cephalochordate *Branchiostoma floridae* (Fig. 3). These data suggest that htt in amphioxus acquires functions that are critical for brain development. Htt pro-neurulation function further increased in fishes and mammals, which exhibit the highest pro-rosettes activity (Fig. 3). *Dhtt* was unable to complement the rosetteless phenotype (Fig. 3). Htt pro-neurulation function is therefore peculiar to deuterostomes organisms.

In support of the notion that the htt pro-neurulation function is a chephalochordate innovation, a study by Kauffman and colleagues (2003) reported that htt expression was prevalently non-neural in *Heliocidaris herithrogramma* (sea urchin), in which the nervous system is poorly organized in a nerve ring. In contrast, analysis of htt expression in the tunicate *Halocynthia roretzi* showed that the protein was particularly enriched in the nervous system (as in vertebrates; Kauffman et al. 2003). Consistent with the above results, htt expression during amphioxus development was particularly enriched in the neural tube (Candiani et al. 2007). These data provide further support to the notion that amphioxus htt may play a role in events occurring during neurulation.

The discovery that htt acquired pro-neurulation functions in amphioxus led us to propose that the protein might mark the beginning of the evolution of the nervous system in chordates. Amphioxus occupies a key position in the evolutionary path that leads to the transition from invertebrates to vertebrates and to the acquisition of a more complex nervous system (Holland et al. 2004). The study of key developmental genes in this organism has shed light on the evolution of such vertebrate organs as the brain, kidney, pancreas and pituitary, and of the genetic mechanisms of early embryonic patterning in general (Holland et al. 2004). Htt may therefore be

Fig. 3 Huntingtin pro-neurulation function emerged during deuterostome evolution. The ability of htt to promote in vitro neurulation (neuroepithelial rosettes formation) is subtle in *Dyctiostelium discoideum* (*Dm*), as in lower deuterostomes *Strongylocentrotus purpuratus* (*Sp*) and *Ciona intestinalis* (*Ci*). This function appears in *Branchistoma floridae* (*Bf*) and increases in fish *Danio rerio* (*Dr*), and mammals *Mus Musculus* (*Mm*) and *Homo sapiens* (*Hs*). Htt from *Drosphila melanogaster* (*Dm*) htt does not exhibit pro-rosettes activity. *Bars* represent the quantification of Nestin-positive cells inside neuroepithelial rosettes

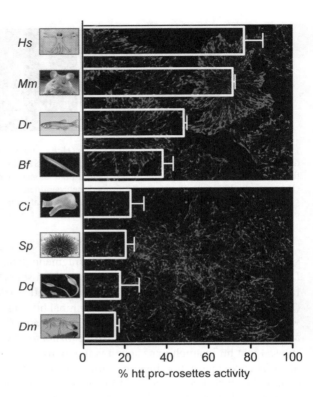

one component of a panel of molecular determinants that drove the transition towards vertebrate nervous system complexity.

ADAM10, A New Target of htt in Neurulation

One of the mechanisms by which htt regulates neurulation involves the cell adhesion protein N-cadherin and its activator disintegrin and metalloproteinase domain-containing protein 10 (ADAM10; Lo Sardo et al. 2012). Previous studies have linked N-cadherin to neural tube formation. The specific expression of N-cadherin in the neural plate following neural induction (Hatta and Takeichi 1986; Detrick et al. 1990; Radice et al. 1997) led to the hypothesis that this adhesion molecule might be specifically required for neural tube morphogenesis. Several studies have provided support to this hypothesis. Misexpression of N-cadherin in the non-neural ectoderm in *Xenopus* affected neural tube size and organization, suggesting that N-cadherin defined the tissue undergoing neurulation (Detrick et al. 1990; Fujimori et al. 1990). Studies in zebrafish and mouse confirmed that N-cadherin was implicated in the morphogenesis of neural tube by regulating adhesion of neuroepithelial cells and, as recently emerged, also by regulating cell polarity and proliferation (Radice et al. 1997; Lele et al. 2002; Hong and Brewster, 2006; Chalasani and Brewster 2011). Mutations in the N-cadherin gene result in blockage of neural tube formation in the midbrain-hindbrain region and several other neural defects (Lele et al. 2002).

In 2005, a study from the group of P. Saftig at Kiel University showed that the activity of N-cadherin was regulated by the metalloprotease ADAM10 (Reiss et al. 2005). The ADAMs are a family of type I transmembrane proteins that combine features of both cell adhesion molecules and proteinases. They play an important role in the development of the nervous system, where they regulate proliferation, migration, differentiation, axonal growth and myelination (Yang et al. 2006). ADAM10 is the major proteinase responsible for N-cadherin ectodomain cleavage. ADAM10-mediated N-cadherin cleavage induces the generation of a 95 kDa fragment called NTF, that is released extracellularly, and a 40 kDa C-terminal fragment termed CTF1, which can be further processed by a γ-secretase-like activity into a soluble 35 kDa CTF2 (Reiss et al. 2005). The ADAM10-induced N-cadherin cleavage also affects the WNT signalling pathway and β-catenin-mediated gene transcription. Perturbation of this pathway leads to changes in the adhesive behavior of cells (Reiss et al. 2005).

More recent findings have revealed that association of ADAM10 with synapse–associated protein 97 (SAP97) is critical to promoting ADAM10 trafficking to the plasma membrane (Marcello et al. 2007), modulation of synaptic activity and N-cadherin cleavage (Malinverno et al. 2010).

We found that htt binds to ADAM10 and regulates the formation of the SAP97/ADAM10 complex (Lo Sardo et al. 2012). In the absence of htt, formation of the SAP97/ADAM10 complex is increased as well as ADAM10 transport to the plasma membrane and its activation (Lo Sardo et al. 2012). In our ES cell-derived rosette assays, this caused increased N-cadherin cleavage, defects in neuroepithelial cell adhesion and a *rosetteless* phenotype. Genetic and pharmacological inhibition of ADAM10 activity partially rescued the *rosetteless* phenotype in the absence of htt, thus confirming that htt operates through ADAM10 and N-cadherin to drive neurulation (Lo Sardo et al. 2012). Increased levels of cleaved N-cadherin were also detected in the htt-MO zebrafish embryo that showed neural tube organization defects from somite stage 7–8 to the complete formation of the neural tube. Inhibition of ADAM10 activity by treatment with ADAM10 blocker GI254023X improved the overall morphology of the neural tube in htt-MO zebrafish (Lo Sardo et al. 2012), demonstrating that htt also regulates ADAM10 activity in vivo.

There is growing evidence implicating a role for ADAM10 in the development of the nervous system. In addition to N-cadherin, ADAM10 is responsible for the shedding of several cell surface proteins such as Ephrins, amyloid precursor protein and Notch that are all involved in brain development (Yang et al. 2006). It may thus be possible that, in addition to N-cadherin, htt regulates the activity of other ADAM10 substrates during brain development.

Conclusions

Understanding the molecules and pathways responsible for the evolution and development of the brain is one of the most fascinating endeavors facing modern science. We hypothesize that htt and its polyQ are potential partners in the complex series of events that led to the formation of a centralized nervous system in deuterostomes. Htt's origin dates back to before the protostome-deuterostome divergence. However, the polyQ in the NH_2-terminus of htt is a recent acquisition, as it appeared and it expanded during deuterostome evolution in organisms with increased nervous system complexity. The appearance of the polyQ in htt of echinoderms and its stabilization in chordates may have implications for the evolutionary path that has led to the transition from invertebrates to vertebrates and to the acquisition of morphologically more complex nervous systems.

Our studies also propose ADAM10 as a novel target in HD. ADAM10 is expressed at high levels in the adult brain. By acting on a large panel of synaptic substrates, including cell adhesion molecules, receptors and ion channels, ADAM10 is critical for synapse structure and function (Pruessmeyer and Ludwig 2009). ADAM10 activity and N-cadherin cleavage are increased in htt-deficient adult brain tissues, which indicates that htt's control over ADAM10-mediated signalling is conserved in the mammalian adult brain (Lo Sardo et al. 2012). This finding suggests that htt may control cell-cell adhesion in the mature brain, perhaps to promote neuronal plasticity or synapse remodelling. HD is characterized by dysfunctions in synaptic circuitries (Cepeda et al. 2007; Fan and Raymond 2007). We believe that the search for effective strategies to help restore the activity of the cortico-striatal synapse (regarding structure, function, and plasticity) is an important goal of HD research that requires further studies.

References

Abranches E, Silva M, Pradier L, Schulz H, Hummel O, Henrique D, Bekman E (2009) Neural differentiation of embryonic stem cells in vitro: a road map to neurogenesis in the embryo. PLoS One 4:e6286

Andrade MA, Bork P (1995) HEAT repeats in the Huntington's disease protein. Nat Genet 11:115–116

Arthur W (2004) The effect of development on the direction of evolution: toward a twenty-first century consensus. Evol Dev 6:282–288

Candiani S, Pestarino M, Cattaneo E, Tartari M (2007) Characterization, developmental expression and evolutionary features of the huntingtin gene in the amphioxus Branchiostoma floridae. BMC Dev Biol 7:127

Cattaneo E, Zuccato C, Tartari M (2005) Normal huntingtin function: an alternative approach to Huntington's disease. Nat Rev Neurosci 6:919–930

Cepeda C, Wu N, Andre VM, Cummings DM, Levine MS (2007) The corticostriatal pathway in Huntington's disease. Prog Neurobiol 81:253–271

Chalasani K, Brewster RM (2011) N-cadherin-mediated cell adhesion restricts cell proliferation in the dorsal neural tube. Mol Biol Cell 22:1505–1515

Clabough EB, Zeitlin SO (2006) Deletion of the triplet repeat encoding polyglutamine within the mouse Huntington's disease gene results in subtle behavioral/motor phenotypes in vivo and elevated levels of ATP with cellular senescence in vitro. Hum Mol Genet 15:607–623

Detrick RJ, Dickey D, Kintner CR (1990) The effects of N-cadherin misexpression on morphogenesis in Xenopus embryos. Neuron 4:493–506

Elkabetz Y, Panagiotakos G, Al Shamy G, Socci ND, Tabar V, Studer L (2008) Human ES cell-derived neural rosettes reveal a functionally distinct early neural stem cell stage. Genes Dev 22:152–165

Fan MM, Raymond LA (2007) N-methyl-D-aspartate (NMDA) receptor function and excitotoxicity in Huntington's disease. Prog Neurobiol 81:272–293

Fondon JW 3rd, Hammock EA, Hannan AJ, King DG (2008) Simple sequence repeats: genetic modulators of brain function and behavior. Trends Neurosci 31:328–334

Fujimori T, Miyatani S, Takeichi M (1990) Ectopic expression of N-cadherin perturbs histogenesis in Xenopus embryos. Development 110:97–104

Gissi C, Pesole G, Cattaneo E, Tartari M (2006) Huntingtin gene evolution in Chordata and its peculiar features in the ascidian Ciona genus. BMC Genomics 7:288

Godin JD, Colombo K, Molina-Calavita M, Keryer G, Zala D, Charrin BC, Dietrich P, Volvert ML, Guillemot F, Dragatsis I, Bellaiche Y, Saudou F, Nguyen L, Humbert S (2010) Huntingtin is required for mitotic spindle orientation and mammalian neurogenesis. Neuron 67:392–406

Gunawardena S, Her LS, Brusch RG, Laymon RA, Niesman IR, Gordesky-Gold B, Sintasath L, Bonini NM, Goldstein LS (2003) Disruption of axonal transport by loss of huntingtin or expression of pathogenic polyQ proteins in Drosophila. Neuron 40:25–40

Hatta K, Takeichi M (1986) Expression of N-cadherin adhesion molecules associated with early morphogenetic events in chick development. Nature 320:447–449

Henshall TL, Tucker B, Lumsden AL, Nornes S, Lardelli MT, Richards RI (2009) Selective neuronal requirement for huntingtin in the developing zebrafish. Hum Mol Genet 18:4830–4842

Holland LZ, Laudet V, Schubert M (2004) The chordate amphioxus: an emerging model organism for developmental biology. Cell Mol Life Sci 61:2290–2308

Hong E, Brewster R (2006) N-cadherin is required for the polarized cell behaviors that drive neurulation in the zebrafish. Development 133:3895–3905

Hughes AL, Friedman R (2005) Loss of ancestral genes in the genomic evolution of Ciona intestinalis. Evol Dev 7:196–200

Kashi Y, King D, Soller M (1997) Simple sequence repeats as a source of quantitative genetic variation. Trends Genet 13:74–78

Kauffman JS, Zinovyeva A, Yagi K, Makabe KW, Raff RA (2003) Neural expression of the Huntington's disease gene as a chordate evolutionary novelty. J Exp Zoolog B Mol Dev Evol 297:57–64

Koren Z, Trifonov EN (2011) Role of everlasting triplet expansions in protein evolution. J Mol Evol 72:232–239

Lele Z, Folchert A, Concha M, Rauch GJ, Geisler R, Rosa F, Wilson SW, Hammerschmidt M, Bally-Cuif L (2002) Parachute/n-cadherin is required for morphogenesis and maintained integrity of the zebrafish neural tube. Development 129:3281–3294

Li Z, Karlovich CA, Fish MP, Scott MP, Myers RM (1999) A putative Drosophila homolog of the Huntington's disease gene. Hum Mol Genet 8:1807–1815

Lo Sardo V, Zuccato C, Gaudenzi G, Vitali B, Ramos C, Tartari M, Myre MA, Walker JA, Pistocchi A, Conti L, Valenza M, Drung B, Schmidt B, Gusella J, Zeitlin S, Cotelli F, Cattaneo E (2012) An evolutionary recent neuroepithelial cell adhesion function of huntingtin implicates ADAM10-Ncadherin. Nat Neurosci 15:713–721

MacDonald ME (2003) Huntingtin: alive and well and working in middle management. Sci STKE 2003:pe48

MacDonald ME, Ambrose CM, Duyao MP, Richard H, Myer RH, Lin C, Srinidhi L, Barnes G, Taylor SA, James M, Groot N, MacFarlane H, Jenkins B, Anderson MA, Wexler NS, Bates

JFGG, Baxendale S, Hummerich H, Kirby S, North M, Youngman S, Mott R, Zehetner G, Sedlacek Z, Poustka A, Frischauf AM, Lehrach H, Buckler AJ, Church D, Doucette-Stamm L, O'Donovan MC, Riba-Ramirez L, Shah M, Stanton VP, Strobel SA, Draths KM, Wales JL, Dervan P, Altherr DEHM, Shiang R, Thompson L, Fielder T, Wasmuth JJ, Tagle D, Valdes J, Elmer L, Allard M, Castilla L, Swaroop M, Blanchard K, Collins FS, Snell R, Holloway T, Gillespie K, Datson N, Shaw D, Harper PS (1993) A novel gene containing a trinucleotide repeat that is expanded and unstable on Huntington's disease chromosomes. Cell 72:971–983

Malinverno M, Carta M, Epis R, Marcello E, Verpelli C, Cattabeni F, Sala C, Mulle C, Di Luca M, Gardoni F (2010) Synaptic localization and activity of ADAM10 regulate excitatory synapses through N-cadherin cleavage. J Neurosci 30:16343–16355

Marcello E, Gardoni F, Mauceri D, Romorini S, Jeromin A, Epis R, Borroni B, Cattabeni F, Sala C, Padovani A, Di Luca M (2007) Synapse-associated protein-97 mediates alpha-secretase ADAM10 trafficking and promotes its activity. J Neurosci 27:1682–1691

Muhlau M, Winkelmann J, Rujescu D, Giegling I, Koutsouleris N, Gaser C, Arsic M, Weindl A, Reiser M, Meisenzahl EM (2012) Variation within the Huntington's disease gene influences normal brain structure. PLoS One 7:e29809

Myre MA, Lumsden AL, Thompson MN, Wasco W, MacDonald ME, Gusella JF (2011) Deficiency of huntingtin has pleiotropic effects in the social amoeba Dictyostelium discoideum. PLoS Genet 7:e1002052

Palidwor GA, Shcherbinin S, Huska MR, Rasko T, Stelzl U, Arumughan A, Foulle R, Porras P, Sanchez-Pulido L, Wanker EE, Andrade-Navarro MA (2009) Detection of alpha-rod protein repeats using a neural network and application to huntingtin. PLoS Comput Biol 5:e1000304

Perutz MF, Johnson T, Suzuki M, Finch JT (1994) Glutamine repeats as polar zippers: their possible role in inherited neurodegenerative diseases. Proc Natl Acad Sci USA 91:5355–5358

Pruessmeyer J, Ludwig A (2009) The good, the bad and the ugly substrates for ADAM10 and ADAM17 in brain pathology, inflammation and cancer. Semin Cell Dev Biol 20:164–174

Radice GL, Rayburn H, Matsunami H, Knudsen KA, Takeichi M, Hynes RO (1997) Developmental defects in mouse embryos lacking N-cadherin. Dev Biol 181:64–78

Ramazzotti M, Monsellier E, Kamoun C, Degl'Innocenti D, Melki R (2012) Polyglutamine repeats are associated to specific sequence biases that are conserved among eukaryotes. PLoS One 7:e30824

Reiner A, Del Mar N, Meade CA, Yang H, Dragatsis I, Zeitlin S, Goldowitz D (2001) Neurons lacking huntingtin differentially colonize brain and survive in chimeric mice. J Neurosci 21:7608–7619

Reiss K, Maretzky T, Ludwig A, Tousseyn T, de Strooper B, Hartmann D, Saftig P (2005) ADAM10 cleavage of N-cadherin and regulation of cell-cell adhesion and beta-catenin nuclear signalling. EMBO J 24:742–752

Rigamonti D, Bauer JH, De-Fraja C, Conti L, Sipione S, Sciorati C, Clementi E, Hackam A, Hayden MR, Li Y, Cooper JK, Ross CA, Govoni S, Vincenz C, Cattaneo E (2000) Wild-type huntingtin protects from apoptosis upstream of caspase-3. J Neurosci 20:3705–3713

Rubinsztein DC, Amos W, Leggo J, Goodburn S, Ramesar RS, Old J, Bontrop R, McMahon R, Barton DE, Ferguson-Smith MA (1994) Mutational bias provides a model for the evolution of Huntington's disease and predicts a general increase in disease prevalence. Nat Genet 7:525–530

Ruden DM, Garfinkel MD, Xiao L, Lu X (2005) Expansions and contractions and the "biased embryos" hypothesis for rapid morphological evolution. Curr Genomics 6:145–155

Sathasivam K, Baxendale S, Mangiarini L, Bertaux F, Hetherington C, Kanazawa I, Lehrach H, Bates GP (1997) Aberrant processing of the Fugu HD (FrHD) mRNA in mouse cells and in transgenic mice. Hum Mol Genet 6:2141–2149

Schaefer MH, Wanker EE, Andrade-Navarro MA (2012) Evolution and function of CAG/polyglutamine repeats in protein-protein interaction networks. Nucleic Acids Res 40:4273–4287

Steffan JS, Agrawal N, Pallos J, Rockabrand E, Trotman LC, Slepko N, Illes K, Lukacsovich T, Zhu YZ, Cattaneo E, Pandolfi PP, Thompson LM, Marsh JL (2004) SUMO modification of Huntingtin and Huntington's disease pathology. Science 304:100–104

Tartari M, Gissi C, Lo Sardo V, Zuccato C, Picardi E, Pesole G, Cattaneo E (2008) Phylogenetic comparison of huntingtin homologues reveals the appearance of a primitive polyQ in sea urchin. Mol Biol Evol 25:330–338

Wang Y, Steimle PA, Ren Y, Ross CA, Robinson DN, Egelhoff TT, Sesaki H, Iijima M (2011) Dictyostelium huntingtin controls chemotaxis and cytokinesis through the regulation of myosin II phosphorylation. Mol Biol Cell 22:2270–2281

Whan V, Hobbs M, McWilliam S, Lynn DJ, Lutzow YS, Khatkar M, Barendse W, Raadsma H, Tellam RL (2010) Bovine proteins containing poly-glutamine repeats are often polymorphic and enriched for components of transcriptional regulatory complexes. BMC Genomics 11:654

White JK, Auerbach W, Duyao MP, Vonsattel JP, Gusella JF, Joyner AL, MacDonald ME (1997) Huntingtin is required for neurogenesis and is not impaired by the Huntington's disease CAG expansion. Nat Genet 17:404–410

Yang P, Baker KA, Hagg T (2006) The ADAMs family: coordinators of nervous system development, plasticity and repair. Prog Neurobiol 79:73–94

Ying QL, Stavridis M, Griffiths D, Li M, Smith A (2003) Conversion of embryonic stem cells into neuroectodermal precursors in adherent monoculture. Nat Biotechnol 21:183–186

Zhang S, Feany MB, Saraswati S, Littleton JT, Perrimon N (2009) Inactivation of Drosophila Huntingtin affects long-term adult functioning and the pathogenesis of a Huntington's disease model. Dis Model Mech 2:247–266

Zheng Q, Joinnides M (2009) Hunting for the function of Huntingtin. Dis Model Mech 2:199–200

Zuccato C, Tartari M, Crotti A, Goffredo D, Valenza M, Conti L, Cataudella T, Leavitt BR, Hayden MR, Timmusk T, Rigamonti D, Cattaneo E (2003) Huntingtin interacts with REST/NRSF to modulate the transcription of NRSE-controlled neuronal genes. Nat Genet 35:76–83

Zuccato C, Valenza M, Cattaneo E (2010) Molecular mechanisms and potential therapeutical targets in Huntington's disease. Physiol Rev 90:905–981

Human Pluripotent and Multipotent Stem Cells as Tools for Modeling Neurodegeneration

Jerome Mertens, Philipp Koch, and Oliver Brüstle

Abstract Sophisticated protocols for the derivation of defined somatic cell cultures from human pluripotent stem cells have opened fascinating prospects for modeling human diseases in vitro. This is particularly relevant for the study of nervous system disorders, where so far the lack of primary tissue has precluded the development of standardized, cell-based in vitro models. We have recently described the derivation of long-term, self-renewing neuroepithelial stem cells (lt-NES cells) from human pluripotent stem cells. Here we report on how this stable somatic stem cell population can be used to study pathogenic mechanisms underlying Alzheimer's disease (AD) and polyglutamine disorders. Specifically, we demonstrate that human neurons derived from lt-NES cells exhibit the entire machinery required for proteolytic processing of the amyloid precursor protein (APP) and are suitable for studying mutants associated with familial variants of AD as well as pharmaceutical compounds modulating the formation of amyloid beta (Aβ). Using induced pluripotent stem cell-derived lt-NES cells from patients with Machado-Joseph disease as an example, we further show that this cellular model provides experimental access to the molecular events initiating pathological protein aggregation – one of the most common denominators of human neurodegenerative disease.

Introduction

Representing a huge threat for the aging individual as well as for the aging societies of the world, neurodegenerative diseases have become a major focus of biomedical research. Generation and aggregation of misfolded proteins are common

J. Mertens • P. Koch • O. Brüstle (✉)
Institute of Reconstructive Neurobiology, Life & Brain Center, University of Bonn and Hertie Foundation, Sigmund-Freud-Strasse 25, Bonn 53127, Germany
e-mail: brustle@uni-bonn.de

F.H. Gage and Y. Christen (eds.), *Programmed Cells from Basic Neuroscience to Therapy*, Research and Perspectives in Neurosciences 20,
DOI 10.1007/978-3-642-36648-2_6, © Springer-Verlag Berlin Heidelberg 2013

phenomena observed in many neurodegenerative diseases and are considered to be key steps in their pathogenesis. Despite immense efforts, no effective treatment that can stop or reverse progressive neurodegeneration is available yet. Given the inaccessibility of disease-affected human neurons for disease-related research and drug development, such studies have, so far, mostly focused on post-mortem tissue, transgenic animals or non-human cell models. At the same time, it has become clear that many of the underlying pathogenic key mechanisms are specific to the human system and involve intracellular processes characteristic of neurons.

The ability to generate human neurons from human pluripotent stem cells (hPSC) provides a general solution to overcome the limited access to primary tissue and is expected to offer significant advantages for studying pathogenic mechanisms in an authentic cellular context and even in disease- and patient-specific cells (Takahashi et al. 2007). One major prerequisite for the successful biomedical exploitation of human embryonic stem cells (hESC) and induced pluripotent stem cells (iPSC) in studying neurodegenerative diseases is efficient and robust in vitro differentiation protocols for generating the required neuronal subtypes as defined cultures with minimal batch-to-batch variation.

In this chapter, we would like to propose long-term, self-renewing neuroepithelial stem (lt-NES) cells as a particularly suitable cell type for the establishment of hPSC-based models of neurodegenerative disorders (Koch et al. 2009; Falk et al. 2012). Further, we present data that exemplify how this experimental system can be used to elucidate basic cytopathological processes in polyglutamine disorders and Alzheimer's disease (AD; Koch et al. 2011, 2012).

lt-NES Cells as a Tool for hPSC-Based Disease Modeling

Cultures of human neurons generated from hPSC provide fascinating alternatives to explore diseases directly in their respective human target cell types. To harness hPSC for the study of neurological diseases, one important prerequisite is the stable and reproducible production of highly enriched neuronal cultures from different hESC and iPSC lines (Fig. 1). Commonly applied 'run-though' protocols, where hPSC are directly differentiated into mature neuronal cultures, often suffer from non-neural cell contaminants due to incomplete differentiation, the varying neurogenic potential of different hPSC lines (Kim et al. 2011) and batch-to-batch variations, which are inherent to lengthy differentiation protocols. We reasoned that the derivation of a stable intermediate multipotent neural stem cell (NSC) population should minimize such variations in neuronal yield and, once established from a hPSC line of interest, significantly reduce cultivation times and costs for subsequent differentiation runs. We found that such a NSC population could be established by isolating the neural tube-like structures appearing in plated embryoid bodies (EBs) and propagating them in the presence of the growth factors FGF2 and EGF. Inhibition of SMAD signaling by the synergistically acting inhibitors Noggin and SB431542 during hPSC differentiation (Chambers et al. 2009) enables the

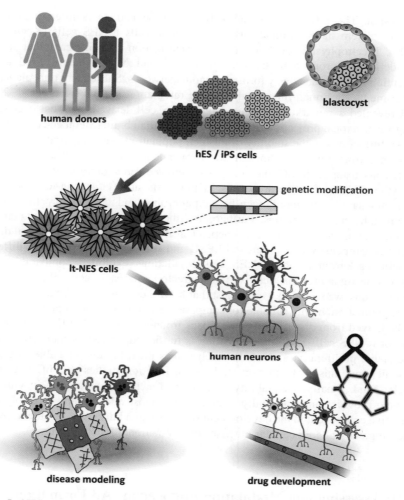

Fig. 1 Lt-NES cells as enabling technology for cell-based disease modeling and drug discovery. The pronounced proliferation potential of lt-NES cells enables continuous and reproducible in vitro production of human neurons from both hESC and iPSC without resorting to the pluripotent stem cell stage. Moreover, genetic modification steps requiring extensive selection can be conducted directly at the NSC stage. The robustness of lt-NES cells makes them particularly suitable for industrial applications such as automated culturing and high throughput screening

generation of such NSC without the need for EB formation. The resulting cells exhibit a characteristic growth pattern with formation of rosette-like structures. They can be expanded for dozens of passages and, upon growth factor withdrawal, always generate the same numbers of neurons and glia. Most importantly, they remain amenable to morphogens such as sonic hedgehog (SHH), FGF8 and retinoic acid, which can be exploited to recruit them into different subtypes such as midbrain dopamine neurons and spinal motoneurons even after extensive expansion. The robustness of this cell population and its neuroepithelium-like growth pattern have prompted us to term these cells lt-NES cells (Koch et al. 2009). Electrophysiological in vitro characterization demonstrated that lt-NES cell-derived neurons expressed functional K^+ and Na^+ channels, fired action potentials and spontaneously formed synaptic networks. Subsequent transplantation experiments confirmed functional synaptic integration in vivo (Koch et al. 2009) and revealed a remarkable potential of lt-NES cell-derived neurons to generate long-range axonal projections in the adult mammalian brain (Steinbeck et al. 2012).

To further assess the stability of this cell population, we compared lt-NES cells generated from nine hPSC lines in two independent laboratories. We found that lt-NES cells from all hESC and iPSC lines showed remarkably similar and stable growth and differentiation properties, including the ability for long-term proliferation, amenability to morphogen-based patterning and competency to generate functionally mature neurons. Furthermore, they showed highly similar neuronal differentiation rates, which were independent from differences in the neural differentiation propensity of the parental hESC and iPSC lines. Despite their different genetic backgrounds, the lt-NES cell populations exhibited comparable gene expression signatures characteristic of neurectodermal stem cells. Based on these observations, we would like to suggest that lt-NES cells might serve as a 'standard' NSC population, highly suitable for comparative studies involving pluripotent stem cells derived from disease and control backgrounds (Falk et al. 2012). An application of lt-NES cells in disease modeling is further supported by their pronounced and stable proliferative potential, which makes them highly amenable to genetic modification, thus enabling gain- and loss-of-function studies on a hESC or iPSC background. Finally, their ability to undergo functional maturation in vitro and in vivo provides an opportunity to address the most relevant and probably most sensitive component in the development of neurological diseases, i.e., the impact of pathologically altered cellular mechanisms on neuronal function (Fig. 1).

AD: Assessing and Modulating Endogenous Aβ Formation in Human Neurons

AD is the most prevalent neurodegenerative disorder. It affects more than one third of the population over the age of 85 and thus represents a huge social and economic problem (Cummings 2008). The extracellular deposition of amyloid beta (Aβ) peptides as insoluble amyloid plaques is a hallmark of the disease. Aβ originates from the amyloid precursor protein (APP) in a sequential proteolytic processing cascade involving beta (β)- and gamma (γ)-secretases (Selkoe 2001). More than 100 mutations in APP and the γ-secretase gene presenilin-1 (PS1) have been shown to cause familial forms of AD (fAD). With all of them directly affecting Aβ generation, they provide a strong argument for the disease-initiating role of Aβ,

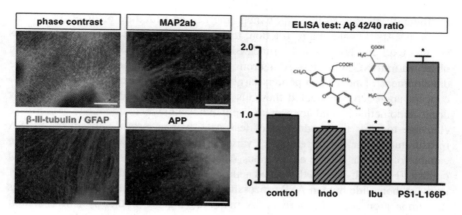

Fig. 2 AD-associated APP processing in authentic human neurons generated from lt-NES cells. Left panel: Upon in vitro differentiation, lt-NES cells give rise to a major fraction of APP-expressing neurons positive for the neuronal marker proteins β-III tubulin and MAP2 as well as a minor fraction of GFAP-positive astrocytes. Right panel: ELISA tests demonstrate that lt-NES cell-derived human neurons efficiently metabolize APP into Aβ and can be used to assess known fAD-associated γ-secretase mutants as well as compounds that impact on the generation of pathological Aβ variants. This proof-of-concept example shows a reduction of the Aβ42/40 ratio upon treatment with the non-steroidal anti-inflammatory drugs indomethacin and ibuprofen and an increased Aβ42/40 ratio in neurons overexpressing the fAD mutant PS1-L166P. *Bar graphs* show mean ± SD. *$P \leq 0.05$ (For further details, see Koch et al. 2012)

the so-called amyloid cascade hypothesis (Hardy and Higgins 1992; De Strooper and Annaert 2010). While the study of Aβ processing represents a key focus of AD-related research, it is limited by two fundamental issues. First, APP processing is highly species-specific, with the human and murine Aβ amino acid sequences differing in >7 %, which strongly affects β-secretase processing (Fung et al. 2004). Secondly, human APP can be expressed in 13 splice variants, whereas only the 695 amino acid-long isoform is specifically expressed in neurons (Hung et al. 1992). Yet, many existing cellular AD models are based on the overexpression of human APP695 in non-neuronal cells. In such models, there is a risk that vastly increased APP levels result in distorted protein sorting, altered subcellular localization and possible artificial effects on downstream actors due to overexerted protein-protein interactions. We have explored whether neurons generated from hESC- and iPSC-derived lt-NES cells provide a more adequate model system for studying APP processing (Fig. 2). As in most cell lines, proliferative lt-NES cells mainly express ubiquitous isoforms of APP such as APP751, whereas the neuron-specific APP695 is only marginally expressed. Upon growth factor withdrawal and induction of neuronal differentiation, endogenous expression of APP695 gradually increases at the expense of non-neuronal isoforms and locates into the axonal compartment of the emerging postmitotic neurons. Importantly, neuronal APP695 is efficiently processed into Aβ, which emerged as a mixture of diverse isoforms including Aβ40 as the major fraction, the disease-associated variant Aβ42 and as yet less explored N-terminally truncated Aβ variants. The robust APP processing observed in lt-NES cell-derived neurons makes them an attractive platform for the development of novel compounds targeting the amyloid cascade. In a proof-of-concept approach, we treated the cells with the non-steroidal anti-inflammatory drugs indomethacin and ibuprofen, compounds known to modulate γ-secretase cleavage towards the production of less Aβ42 versus Aβ40. Subsequent ELISA measurements for the different Aβ variants revealed that both compounds were capable of decreasing the endogenous Aβ42/Aβ40 ratio in these human neuronal cultures (Fig. 2). lt-NES cell-derived neurons thus represent an attractive system for studying bona fide human APP processing and for assessing and identifying drugs that impact on secretase activity. A major advantage of this system is that it bypasses the non-physiological expression levels typically found in transgenic models. Especially in the context of the yet unsuccessful drug development for AD, lt-NES cell-derived neuronal cultures may provide a reliable, authentic, and hopefully more predictive platform for preclinical drug validation (Koch et al. 2012).

We also explored the feasibility of this system for studying mutants associated with fAD. Using lentiviral vectors, we overexpressed the aggressive early-onset fAD-associated PS1 mutant L166P in lt-NES cells. Upon growth factor withdrawal, these cells gave rise to PS1-transgenic neuronal cultures. When compared to neurons overexpressing wild-type PS1 or the biochemically non-functional mutant PS1-D385N, the PS1-L166P neurons displayed a strongly increased Aβ42/40 ratio similar to the changes observed in the cerebrospinal fluid of fAD patients (Fig. 2). Interestingly, the Aβ42/40 increase of PS1-L166P proved to be due to a partial loss-of-function in the generation of Aβ40 rather than to an increase in the production of Aβ42. Furthermore, these fAD neurons appeared to be completely resistant to γ-secretase modulating compounds that were efficient in the context of non-transgenic neurons and neurons overexpressing wild type PS1 (Koch et al. 2012).

In principal, this approach can also be extended to lt-NES cells derived from patient-specific iPSC. Such iPSC lines have become increasingly available, including lines from AD patients with point mutations in the PS1 and PS2 genes (Yagi et al. 2011) or duplications of the APP gene on chromosome 21 (Israel et al. 2012). It will also be interesting to subject patient-specific NSC and neurons to a lentiviral overexpression paradigm to study possible additive effects of pertinent disease-associated proteins such as variants of apolipoproteins or the relationship between an increased phosphorylation of tau and Aβ toxicity. At the same time, lentiviral overexpression systems might enable rescue experiments in iPSC-derived neurons from AD patients.

Machado-Joseph Disease: The Excitation–Aggregation Connection

While altered APP and tau processing represent key pathogenic components underlying AD, other neurodegenerative diseases are characterized by an extension of polyglutamine repeats in various proteins. These so-called polyglutamine diseases comprise different neurological disorders, including Huntington's disease (HD), various spinocerebellar ataxias, spinal and bulbar muscular atrophy (SBMA), also known as Kennedy disease (KD), and dentatorubropallidoluysian atrophy (DRPLA), which mainly occurs in the Japanese population. The unifying feature of all polyglutamine diseases is the abnormal expansion of trinucleotide CAG repeats within an affected gene. Trinucleotide repeat-containing genes are unstable, with the CAG repeats frequently expanding across intergenerational transmission. Furthermore, the repeat length appears to determine the age of onset as well as the severity of disease progression in all polyglutamine diseases (Ha and Fung 2012). Pathologically elongated CAG repeats are translated into polyglutamine stretches within the respective gene products, which in turn are considered to lead to an aberrant protein conformation that causes a pathogenic cascade leading to neuronal dysfunction (La Spada and Taylor 2010).

We decided to harness our lt-NES cell system to generate an iPSC-based model for the late-onset polyglutamine disease Machado Joseph disease (MJD; syn. spinocerebellar ataxia type 3; SCA3), which represents the most common, dominantly inherited ataxia. The disease is caused by abnormal CAG expansions in the MJD1 gene, which translate into extended polyglutamine stretches within the ataxin-3 protein. The latter are thought to lead to conformational changes resulting in protein dysfunction and aggregation. The disease results in the degeneration of neurons in the brainstem and other regions (Costa Mdo and Paulson 2012) and is characterized by a progressive decrease in movement control and orientation skills.

To investigate the early steps of the disease in neurons, patient-specific iPSC were generated from skin biopsies from MJD patients and healthy siblings (Koch et al. 2011). Following a comprehensive characterization of the pluripotent stem cells, lt-NES cells were derived from each iPSC line as a stable source for the generation of mature neuronal cultures. One of the earliest steps in ataxin-3 processing is proteolytic cleavage, resulting in a C-terminal fragment containing the extended polyglutamine stretch. While the protein features predict caspase and calpain cleavage sites, the precise nature of cleavage in humans is, so far, unknown (Tarlac and Storey 2003). Since these enzymes depend on calcium, and intracellular calcium levels strongly fluctuate with neuronal activity, we became interested in whether formation of ataxin-3 aggregates depends on neuronal activity. To address this question, we stimulated electrophysiologically mature neuronal cultures with the neurotransmitter glutamate, which led to increased Ca^{2+} transients within the cells. Protein lysates generated after such stimulation indeed showed induction of ataxin-3 cleavage in iPSC-derived neurons from MJD patients and healthy controls. However, only MJD neurons, but not neurons from healthy donors, showed the

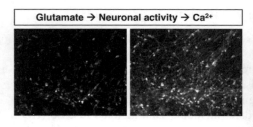

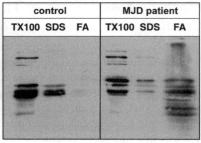

Fig. 3 Lt-NES cells in iPSC-based modeling of the polyglutamine disorder MJD. Differentiated lt-NES cells yield functional neurons that can be used to address pathogenic mechanisms associated with neuronal function. *Left panel*: Stimulation of lt-NES cell-derived neurons with glutamate results in an increase in intracellular Ca^{2+}. *Right panel*: This activity-dependent increase of intracellular Ca^{2+} induces calpain-mediated cleavage of ataxin-3, which initiates the formation of SDS-insoluble ataxin-3-containing microaggregates detected in the formic acid (FA) fraction of neuronal culture lysates from MJD patients but not healthy controls

formation of SDS-insoluble microaggregates consisting of polyglutamine-containing ataxin-3 fragments (Fig. 3). Subsequent experiments involving specific pharmacological blockade of different proteases revealed that microaggregate formation could be prevented with calpain inhibitors but not caspase inhibitors, indicating that calpain-mediated cleavage is the key mechanism in early proteolytic processing of ataxin-3. Microaggregate formation could also be inhibited by capturing extracellular Ca^{2+} via EDTA or glutamate receptor blockade. Most interestingly, the generation of microaggregates was also abolished upon blockade of voltage-gated Na^+ channels with tetrodotoxin, indicating that depolarization-induced activation of voltage-gated Ca^{2+} channels was required for an intracellular Ca^{2+} increase sufficient to trigger calpain-mediated cleavage of ataxin-3. Taken together, these results also provide an explanation of why MJD is a disorder that specifically affects neurons. Accordingly, when the experiment was repeated using patient fibroblasts, iPSC or astrocytes, no aggregation of ataxin-3 could be detected. It is an intriguing question whether the same basic mechanisms we observed in our model also apply to the situation in vivo, and whether functional activity determines the distribution of neuronal degeneration in the brain of MJD patients. Unlike disorders such as Parkinson's disease and HD, which result in predominant damage of specific neurotransmitter subtypes, the neurons affected in MJD lack such a common denominator. However, the neuronal populations affected by MJD are part of a highly and continuously active circuitry involved in the control of movement and posture, which makes an activity-dependent mode of disease an attractive though unproven hypothesis. Remarkably, no signs of neuronal degeneration could be detected upon induction of ataxin-containing microaggregates in our cell system, indicating that iPSC-based disease modeling provides a window for studying early pathogenic cellular mechanisms of late-onset neurodegenerative diseases long before they result in neurodegeneration and classic textbook pathology associated with the end stage of the disease, i.e., mechanisms, that might be particularly attractive for pharmaceutical intervention targeted at disease progression.

Conclusion

As a stable, intermediate stem cell population for the generation of enriched neuronal cultures from hESC and iPSC, lt-NES cells stand out as a reliable and time-saving cell system for the generation of human neuronal disease models. Neurons derived from lt-NES cells closely resemble human neurons in vivo with respect to protein expression patterns, neuron-specific protein metabolism and electrophysiological functionality. As such they represent an interesting tool for deciphering complex pathological pathways that are exclusive to human neurons. The proliferative capacity and robustness of lt-NES cells further allow for the introduction of genetic modifications directly at the NSC stage, yielding, e.g., human neurons carrying disease-related mutations or gene-corrected patient-specific cells for therapeutic modeling. The robustness of lt-NES cells also makes them an attractive cell source for industrialization, including bioreactor-based scale-up, automated culture and differentiation as well as scale-out and high throughput screening in industrial drug discovery.

References

Chambers S, Fasano CA, Papapetrou EP, Tomishima M, Sadelain M, Studer L (2009) Highly efficient neural conversion of human ES and iPS cells by dual inhibition of SMAD signaling. Nat Biotechnol 27:275–280

Costa Mdo C, Paulson HL (2012) Toward understanding Machado-Joseph disease. Prog Neurobiol 97:239–257

Cummings JL (2008) The black book of Alzheimer's disease, Part 1. Primary Psychiat 15:6676

De Strooper B, Annaert W (2010) Novel research horizons for presenilins and gamma-secretases in cell biology and disease. Annu Rev Cell Dev Biol 26:235–260

Falk A, Koch P, Kesavan J, Takashima Y, Ladewig J, Alexander M, Wiskow O, Tailor J, Trotter M, Pollard S, Smith A, Brüstle O (2012) Capture of neuroepithelial-like stem cells from pluripotent stem cells provides a versatile system for in vitro production of human neurons. PLoS One 7:e29597

Fung J, Frost D, Chakrabartty A, McLaurin J (2004) Interaction of human and mouse Abeta peptides. J Neurochem 91:1398–1403

Ha AD, Fung VS (2012) Huntington's disease. Curr Opin Neurol 25:491–498

Hardy JA, Higgins GA (1992) Alzheimer's disease: the amyloid cascade hypothesis. Science 256:184–185

Hung AY, Koo EH, Haass C, Selkoe DJ (1992) Increased expression of beta-amyloid precursor protein during neuronal differentiation is not accompanied by secretory cleavage. Proc Natl Acad Sci USA 89:9439–9443

Israel MA, Yuan SH, Bardy C, Reyna SM, Mu Y, Herrera C, Hefferan MP, Van Gorp S, Nazor KL, Boscolo FS, Carson CT, Laurent LC, Marsala M, Gage FH, Remes AM, Koo EH, Goldstein LS

(2012) Probing sporadic and familial Alzheimer's disease using induced pluripotent stem cells. Nature 482:216–220

Kim JE, O'Sullivan ML, Sanchez CA, Hwang M, Israel MA, Brennand K, Deerinck TJ, Goldstein LS, Gage FH, Ellisman MH, Ghosh A (2011) Investigating synapse formation and function using human pluripotent stem cell-derived neurons. Proc Natl Acad Sci USA 108:3005–3010

Koch P, Opitz T, Steinbeck JA, Ladewig J, Brüstle O (2009) A rosette-type, self-renewing human ES cell-derived neural stem cell with potential for in vitro instruction and synaptic integration. Proc Natl Acad Sci USA 106:3225–3230

Koch P, Breuer P, Peitz M, Jungverdorben J, Kesavan J, Poppe D, Doerr J, Ladewig J, Mertens J, Tüting T, Hoffmann P, Klockgether T, Evert BO, Wüllner U, Brüstle O (2011) Excitation-induced ataxin-3 aggregation in neurons from patients with Machado-Joseph disease. Nature 480:543–546

Koch P, Tamboli IY, Mertens J, Wunderlich P, Ladewig J, Stuber K, Esselmann H, Wiltfang J, Brüstle O, Walter J (2012) Presenilin-1 L166P mutant human pluripotent stem cell-derived neurons exhibit partial loss of gamma-secretase activity in endogenous amyloid-beta generation. Am J Pathol 180:2404–2416

La Spada AR, Taylor JP (2010) Repeat expansion disease: progress and puzzles in disease pathogenesis. Nat Rev Genet 11:247–258

Selkoe DJ (2001) Alzheimer's disease: genes, proteins, and therapy. Physiol Rev 81:741–766

Steinbeck JA, Koch P, Derouiche A, Brüstle O (2012) Human embryonic stem cell-derived neurons establish region-specific, long-range projections in the adult brain. Cell Mol Life Sci 69:461–470

Takahashi K, Tanabe K, Ohnuki M, Narita M, Ichisaka T, Tomoda K, Yamanaka S (2007) Induction of pluripotent stem cells from adult human fibroblasts by defined factors. Cell 131:861–872

Tarlac V, Storey E (2003) Role of proteolysis in polyglutamine disorders. J Neurosci Res 74:406–416

Yagi T, Ito D, Okada Y, Akamatsu W, Nihei Y, Yoshizaki T, Yamanaka S, Okano H, Suzuki N (2011) Modeling familial Alzheimer's disease with induced pluripotent stem cells. Hum Mol Genet 23:4530–4539

Human Stem Cell Approaches to Understanding and Treating Alzheimer's Disease

Lawrence S.B. Goldstein

Abstract This chapter describes how human induced pluripotent stem cell (hIPSC) technologies might be used to study Alzheimer's disease. I argue that significant mechanistic and therapeutic insights may emerge regarding familial Alzheimer's disease and also sporadic Alzheimer's disease by using hIPSC methods. Uniting stem cell, bioengineering, and genetic/genomic technologies may provide a uniquely powerful platform for understanding and treating this terrible human disease.

Introduction

Alzheimer's disease (AD) poses an enormous unsolved scientific and medical problem. AD is very common, with 10 % of people over the age of 65 and 50 % of people over the age of 85 estimated to be afflicted in the United States (Mebane-Sims and Association 2009). There are no disease-altering therapeutics available and, while there are many possible therapies in the pipeline, there is no guarantee that they will be successful. Thus, while the current social burden of the disease is enormous, it will grow substantially in the coming years unless truly effective approaches to disease management are developed and implemented. While it is sometimes thought that this disease only affects the very elderly, in fact many people develop AD at relatively young ages. In addition, many people in their 60s, 70s, and 80s can continue to be very productive if they do not develop AD, and so the emotional and economic impact of AD on families and society is substantial. It is important to remember that if 50 % of people over the age of 85 have AD then 50 % of families with a family member over the age of 85 are coping with the toll of

L.S.B. Goldstein (✉)
Department of Cellular and Molecular Medicine and Department of Neurosciences, Sanford Consortium for Regenerative Medicine and UCSD School of Medicine, 9500 Gilman Drive, La Jolla, CA 92093-0695, USA
e-mail: lgoldstein@ucsd.edu

F.H. Gage and Y. Christen (eds.), *Programmed Cells from Basic Neuroscience to Therapy*, Research and Perspectives in Neurosciences 20, DOI 10.1007/978-3-642-36648-2_7, © Springer-Verlag Berlin Heidelberg 2013 this disease, and hence the impact on much younger people is arguably greater than the impact on the afflicted family member.

Thus far, most work on human subjects with AD has been observational, postmortem, or has worked with human cell types that are not afflicted by disease. This work has established the primary pathology that is classically used to define AD: the amyloid plaques formed primarily from the Aβ peptides generated by proteolytic cleavage of the amyloid precursor protein (APP), and the neurofibrillary tangles (NFT) generated by phosphorylation and aggregation of tau protein (Hardy and Selkoe 2002). Neither of these pathologies correlates sufficiently well with AD cognitive symptoms to fully explain the disease. In fact, considerable evidence suggests that synaptic loss in key regions of the brain may be the best correlate of cognitive loss and almost certainly precedes the neuronal death that is commonly seen (Selkoe 2002).

Most human AD is "sporadic," which means that a combination of genetic predisposition factors and environment conspires in an unknown way to generate the disease. However, there are rare, simple dominant Mendelian forms of AD that are referred to as familial AD (FAD; Bertram and Tanzi 2008, 2009). FAD is caused by mutations in one of three genes in humans. The gene encoding APP will cause FAD when point mutations in and around the region encoding a transmembrane domain are present or when there is a duplication of the gene, i.e., an increase from the normal two doses of the APP gene to three doses is sufficient to cause aggressive early onset AD. Mutations in the presenilin 1 or presenilin 2 genes can also cause aggressive early-onset FAD.

The identification of genes and mutations that cause FAD enabled the generation of mouse models that generate amyloid plaques. These models have been helpful in understanding the biochemical pathways leading to amyloid plaque formation and have contributed to an understanding of factors that promote or inhibit the formation of plaques. A surprise finding from these models was that they did not form NFT and that synaptic and neuronal loss was modest relative to expectations predicted by the dominant amyloid cascade model. The addition of mutant tau genes has apparently improved the fidelity with which these models mimic human AD, but it also highlights the fact that human proteins and mouse proteins behave differently, with tau potentially defining the tip of the iceberg of important human-mouse differences. Thus, there is great interest in, and hope for, the development of true human models of AD that will augment the search for mechanistic understanding and effective therapies.

Human IPSC Models of FAD

The recent development of effective reprogramming technologies allows the generation of human IPSCs (hIPSCs) that contain the genomes of any person if an appropriate cell sample is available (Takahashi et al. 2007). For FAD, a number of papers (including from my own lab) using this novel technology to study AD have been recently published. In our work, we took biopsies from two patients carrying an FAD APP duplication that caused dominant early onset FAD in a Finnish family. Fibroblasts from these biopsies were then used to generate multiple (at least three) independent and well-behaved hIPSC lines. We then took advantage of our recently developed FACS-based purification methods to purify neuronal stem cells and neurons derived from each independent hIPSC line and probe for a number of AD phenotypes (Israel et al. 2012).

First, as expected for a FAD APP duplication, neurons carrying this mutation generated elevated levels of secretion of Aβ peptide. Second, surprisingly, FAD APP duplication neurons generated aberrant activation of the GSK3 kinase and elevated phosphorylation of a tau protein residue that is thought to be pathological. Third, experiments with drugs that inhibit different steps in the proteolytic processing of APP suggested that the AD-related biochemical phenotypes seen in these neurons were not driven by the Aβ peptide but were instead driven by a transmembrane C-terminal fragment of APP. Fourth, FAD APP duplication neurons exhibited enlarged endosomes, comparable to those found in postmortem materials and in APP duplication fibroblasts. A strength of these studies is that purified neurons could be used to observe typical AD phenotypes in short-term culture experiments, opening the way to more extensive mechanistic studies and potentially therapy-development projects. The weaknesses of these studies include a mixed neurotransmitter population of neurons, the short-term nature of the cultures, and the lack of glia, which are crucial for many normal neuronal functions. Clearly, there is more work to be done.

Recent independent experiments confirmed and extended our studies in important ways. This new work used a differentiation system that appears to recapitulate differentiation of cortical glutamatergic neurons over a longer period than that used in our studies (Shi et al. 2012b). Neurons carrying a trisomy 21 that causes Down's syndrome were found to also exhibit elevated Aβ secretion and perhaps deposition into plaque-like aggregates (Shi et al. 2012a). Elevated phosphorylation and pathological relocalization of tau was also reported. Importantly, two very different differentiation and analysis approaches, one using highly purified neurons of diverse neurotransmitter type and the other using less purified neurons but with an apparently more uniform neurotransmitter type, generated similar conclusions supporting the premise that hIPSC-derived neurons will be useful for observing phenotypes and potentially for testing mechanisms and drugs.

Further explorations of hIPSC lines carrying presenilin 1 mutations have also been recently published. In one instance, neurons carrying presenilin 1 mutations were generated by direct induction using transcription factors introduced into mutation-bearing fibroblasts (Qiang et al. 2011). In this case, altered Aβ production typical of presenilin 1 mutations was found. Surprisingly, endosomal enlargement was also found. This phenotype was not previously found to be present in postmortem material from human patients carrying presenilin 1 mutations, although lysosomal abnormalities were found in postmortem material from human patients carrying presenilin 1 mutations. Perhaps the neurons in culture are revealing a variation or precursor of the post-mortem phenotype, which should clearly be probed further. Other papers used reprogramming to generate presenilin 1-carrying hIPSC lines that were in turn used to generate mutation-carrying neurons (Yagi et al. 2011; Yahata et al. 2011). These neurons also exhibited altered Aβ production; no other significant phenotypes have yet been reported. Another report described intriguing studies of Aβ-modulating drugs (Yahata et al. 2011). Further work comparing the various phenotypes in neurons of the various FAD mutations and searches for additional informative phenotypes and mechanisms and therapeutic interventions are in progress in multiple labs.

A number of important questions remain to be asked in these systems, including (1) to what extent are phenotypes neuron-autonomous and are they modified to differing extents by astrocytes carrying different APOE alleles? (2) to what extent does genetic background modify phenotypes caused by APP duplication or APP and presenilin point mutations? (3) does long-term culture in any of these systems lead to the formation of NFTs, synaptic loss, or cell death? and (4) can these systems be used to develop unique insights that could not or have not yet been obtained in animal models of AD?

The Problem of Genetic Background and Sporadic AD

For FAD, issues surrounding the influence of genetic background on the phenotypic consequences of APP or presenilin mutations can ultimately be resolved using gene-targeting technologies. Thus, point mutations can be induced in common genetic backgrounds or point mutations can be repaired to the "wild-type" allele in the same background as a patient carrying a mutation causing FAD. Using hIPSC approaches to probe sporadic AD, however, poses significant, difficult and unique intellectual and technical challenges.

Although sporadic AD is not strictly inherited, the heritability of sporadic AD is high, with estimates that 1/2 to 2/3 of the risk is genetic as based on studies of twins (Gatz et al. 2006). Contributing to this substantial risk are a number of genetic loci identified in genome-wide association studies (GWAS), with APOE being by far the largest contributor (Bertram et al. 2010). Other loci appear to contribute only low levels of average risk, based on GWAS studies. A pessimistic point of view is that such low levels of average risk may make it difficult to usefully study sporadic AD in hIPSC-derived neurons. However, such a view is based in part on the potentially incorrect assumption that, if the average level of risk is low for a given genetic variant, then the contribution of that variant to risk in any given individual would also be low, and so the phenotypic impact of that variant must be low.

There are two theoretical arguments that might promote more optimism about the possibility of using hIPSC methods to study sporadic AD. First, it is possible that genetic variants that cause low levels of average increased risk in the population actually have a distribution of risk levels among individuals. In this scenario, the genetic background may determine the amount of risk a given variant causes in any one person. Thus, some individuals may have a high genetic risk of AD caused by a given genetic variant and some have low risk caused by the same variant, differing only because of the genetic background. An important, and testable, extension of this idea is that, if risk equates to phenotype in culture of glia or neurons, depending on the pathway affected by the variant, some patients would be expected to give rise to hIPSC lines and then neurons or glia that have significant phenotypes.

A second theoretical cause for optimism is that it is possible that the variants that cause an average increased risk of AD in the population all have significant effects on neuronal or glial phenotype, and that might be observed with cells derived from patients and hIPSC lines. In this scenario, the altered phenotype of the neurons or glia might cause varying degrees of impact in an intact brain, depending on other genetic factors that might modulate cell types not examined in culture or other aspects of the physiology of an intact brain that cannot yet be modeled in culture.

It is ultimately possible to test some of the predictions of these varying views and even to imagine testing environmental influences in culture. The difficult catch is that the variants found in GWAS studies are common variants that may be causative or instead may be tightly linked to the causative variants, which themselves could be found in a subset of the population in which they confer a large effect or be some distance away but nonetheless in linkage disequilibrium. Dealing with this set of complications will require significant high-resolution genomic and genetic studies, ultimately requiring gene-targeting studies that might involve large genomic regions. While these studies are feasible with current technology, continued technological improvements will make them much more practical at the scale needed to test genetic contributions to glial and neuronal phenotypes.

In this context, we studied neurons carrying genomes derived from two sporadic AD patients (Israel et al. 2012). We found that one of the two exhibited no obvious AD biochemical phenotypes whereas the other exhibited phenotypes very similar to those seen with a FAD APP duplication. The most rigorous and conservative interpretation of these findings is that there may be genomes in the human population that cause apparently abnormal phenotypes in purified neurons. It is not clear, however, from such a small study whether such genomes predict or drive the development of AD in such individuals. A much larger clinical study needs to be done to ask whether genomes that drive neuronal phenotypes are more common in the sporadic AD population than in non-demented controls.

Some Prospects for the Future

There are several additional areas where neurons derived from reprogrammed stem cells from patients may have a significant impact. First is the search for suppressor or enhancer genes using typical genetic methodologies of mutagenesis, knockdown, or overexpression studies. Second is the search for drug-like molecules that suppress or enhance phenotypes. Both of these approaches can potentially lead to new targets and therapies. A third important avenue is to bring bioengineering technologies to bear that allow cells to be damaged, manipulated or cultured in ways that better mimic the neuronal environments in an intact brain. Chemical agents that cause reactive oxygen and free radical damage, heavy metals, and other suspected environmental insults may be used in combination with the types of genetic and genomic technologies that reprogrammed stem cells can bring to the study of AD. Two additional approaches that may prove useful come from the ability to grow neurons in cultures that allow physical separation between neuronal compartments or perhaps the formation of defined circuits. For example, so-called compartment cultures (Taylor et al. 2005), in which neurons grow axons through narrow channels into an adjacent chamber that is fluidically isolated, allow defined axonal injury as well as culture of glia with axonal but not somatodendritic regions. Similar designs would allow, for example, glutamatergic neurons from one compartment to synapse on cholinergic neurons in another compartment. Taken to a logical engineering extreme, defined circuits – or what might be referred to as a human brain in a dish – might ultimately be constructed for genetic, physiological, and drug studies. Clearly there is much work to be done with this exciting new technology, with the hope that the information that will emerge will help us better understand diseases such as AD and develop more effective therapeutic agents.

References

Bertram L, Tanzi RE (2008) Thirty years of Alzheimer's disease genetics: the implications of systematic meta-analyses. Nat Rev Neurosci 9:768–777

Bertram L, Tanzi RE (2009) Genome-wide association studies in Alzheimer's disease. Hum Mol Genet 18:R137–R145

Bertram L, Lill CM, Tanzi RE (2010) The genetics of Alzheimer disease: back to the future. Neuron 68:270–281

Gatz M, Reynolds CA, Fratiglioni L, Johansson B, Mortimer JA, Berg S, Fiske A, Pedersen NL (2006) Role of genes and environments for explaining Alzheimer disease. Arch Gen Psychiat 63:168–174

Hardy J, Selkoe DJ (2002) The amyloid hypothesis of Alzheimer's disease: progress and problems on the road to therapeutics. Science 297:353–356

Israel MA, Yuan SH, Bardy C, Reyna SM, Mu Y, Herrera C, Hefferan MP, Van Gorp S, Nazor KL, Boscolo FS, Carson CT, Laurent LC, Marsala M, Gage FH, Remes AM, Koo EH, Goldstein LS (2012) Probing sporadic and familial Alzheimer's disease using induced pluripotent stem cells. Nature 482:216–220

Mebane-Sims I, Alzheimer's Association (2009) Alzheimer's disease facts and figures. Alzheimers Dement 5: 234–270

Qiang L, Fujita R, Yamashita T, Angulo S, Rhinn H, Rhee D, Doege C, Chau L, Aubry L, Vanti WB, Moreno H, Abeliovich A (2011) Directed conversion of Alzheimer's disease patient skin fibroblasts into functional neurons. Cell 146:359–371

Selkoe DJ (2002) Alzheimer's disease is a synaptic failure. Science 298:789–791

Shi Y, Kirwan P, Smith J, Maclean G, Orkin SH, Livesey FJ (2012a) A human stem cell model of early Alzheimer's disease pathology in Down syndrome. Sci Transl Med 4:124–129

Shi Y, Kirwan P, Smith J, Robinson HP, Livesey FJ (2012b) Human cerebral cortex development from pluripotent stem cells to functional excitatory synapses. Nat Neurosci 15:477–486

Takahashi K, Tanabe K, Ohnuki M, Narita M, Ichisaka T, Tomoda K, Yamanaka S (2007) Induction of pluripotent stem cells from adult human fibroblasts by defined factors. Cell 131:861–872

Taylor AM, Blurton-Jones M, Rhee SW, Cribbs DH, Cotman CW, Jeon NL (2005) A microfluidic culture platform for CNS axonal injury, regeneration and transport. Nat Meth 2:599–605

Yagi T, Ito D, Okada Y, Akamatsu W, Nihei Y, Yoshizaki T, Yamanaka S, Okano H, Suzuki N (2011) Modeling familial Alzheimer's disease with induced pluripotent stem cells. Hum Mol Genet 20:4530–4539

Yahata N, Asai M, Kitaoka S, Takahashi K, Asaka I, Hioki H, Kaneko T, Maruyama K, Saido TC, Nakahata T, Asada T, Yamanaka S, Iwata N, Inoue H (2011) Anti-Abeta drug screening platform using human iPS cell-derived neurons for the treatment of Alzheimer's disease. PLoS One 6:e25788

Potential of Stem Cell-Derived Motor Neurons for Modeling Amyotrophic Lateral Sclerosis (ALS)

Derek H. Oakley, Gist F. Croft, Hynek Wichterle, and
Christopher E. Henderson

Abstract The human motor system comprises the same basic functional and anatomic categories that have been described in vertebrate model systems. However, almost all of the genetic and molecular information about the development of the motor system and motor neuron subtypes is based on studies in animal models, since human motor neurons have been available only in postmortem samples. With the establishment of human embryonic stem (hES) cells as a research tool, and the demonstration that they could be directed to differentiate into spinal motor neurons, this inaccessibility has changed. Spinal motor neurons are the target of several diseases. As one key example, the progressive paralysis and ultimate death of patients with amyotrophic lateral sclerosis (ALS) reflect changes in motor neuron excitability, selective degeneration of nerve-muscle contacts and finally cell death of vulnerable motor neurons. However, the mechanisms of selective motor neuron degeneration are not well understood and there are no effective therapies. The derivation of induced pluripotent stem (iPS) cells from ALS patients allows human motor neurons and other disease-relevant cell types with the same genetic make-up as the patient to be generated in large numbers. However, before they can be reliably used for mechanistic studies or to establish ALS-relevant screens, they need to be validated as a tool. In this article, we discuss salient aspects of human development and neurodegeneration and consider how they can inform our design of appropriate cell models. In addition, we review our recent data suggesting that

Derek H. Oakley and Gist F. Croft made equal contributions.

D.H. Oakley • G.F. Croft • H. Wichterle • C.E. Henderson (✉)
Project A.L.S./Jenifer Estess Laboratory for Stem Cell Research, Columbia Stem Cell Initiative, Center for Motor Neuron Biology and Disease, Departments of Rehabilitation and Regenerative Medicine, Pathology and Cell Biology, Neurology and Neuroscience, Columbia University Medical Center, New York, NY 10032, USA
e-mail: ch2331@columbia.edu

F.H. Gage and Y. Christen (eds.), *Programmed Cells from Basic Neuroscience to Therapy*, Research and Perspectives in Neurosciences 20,
DOI 10.1007/978-3-642-36648-2_8, © Springer-Verlag Berlin Heidelberg 2013

human iPS cell-derived motor neurons constitute a robust basis for disease modeling and we discuss some of the technological challenges that nevertheless remain to be addressed.

Motor Neuron Degeneration in ALS

ALS is an adult-onset neurodegenerative disease that was first described by Jean-Martin Charcot (Charcot and Joffroy 1869); it is characterized by the selective death of spinal motor neurons and upper motor neurons of the motor cortex. Axonal degeneration and cell loss spread from an initiating motor pool to nearly all others, causing muscle weakness, spasticity and paralysis that lead to death within a few years of diagnosis (Ravits and La Spada 2009). Ninety percent of ALS cases are sporadic, i.e., without known genetic cause; however, 10 % are familial. Mutations in several genes have been linked to ALS, beginning with SOD1 (Rosen et al. 1993) and now including many others (Boillée et al. 2006). For example, mutations in TDP-43 are found in both familial (Gitcho et al. 2008; Van Deerlin et al. 2008) and sporadic (Kabashi et al. 2008) ALS. Even more exciting were recent reports of hexamer expansions in the C9ORF72 gene that are sufficient to explain nearly half of all familial cases of ALS, as well as up to 10 % of sporadic cases (van Es et al. 2009; Dejesus-Hernandez et al. 2011; Renton et al. 2011). Therefore, in addition to a conserved clinical presentation, the molecular distinction between familial and sporadic ALS now appears less watertight than originally believed, further bolstering the hope that mechanistic insights gained from the study of familial gene mutations may have direct significance for the large number of patients with the sporadic form of the disease.

Mutations in the SOD1 gene have formed the basis for rodent models that have provided important insights into the sequence and specifics of pathology, – for example, axonal dieback, misfolded SOD1 and protein aggregation, axonal transport defects, mitochondrial dysfunction, and glutamate excitotoxicity – pointing to key players but not identifying upstream mechanisms that could serve as therapeutic targets. Animal models have also identified, for example, the importance of cell types other than motor neurons, such as astrocytes and microglia, to disease progression (Clement et al. 2003; Boillée et al. 2006). Despite this progress, the disease mechanisms are not well understood and there are no truly effective therapies. This chapter will focus on cell-autonomous, motor neuron-specific aspects of the disease. While these clearly do not fully represent the spectrum of molecular and cellular events that lead to the full pathology, they likely encapsulate the principal therapeutic targets linked to the disease-defining stages of motor neuron degeneration and loss.

Motor Neuron Subtype-Selective Disease Phenotypes

The motor neuron selectivity of cell loss in ALS has historically, and with good reason, occupied much attention. However, while almost all motor neurons degenerate in ALS, certain classes and motor pools show enhanced susceptibility or resistance to degeneration (Kanning et al. 2010). Distal limb projecting or facial motor muscles in bulbar forms are typically affected first, compared to thoracic muscles (Ravits et al. 2007). These clinical findings suggest that limb-innervating motor neurons of the lateral motor column (LMC) are more susceptible to the onset or triggers of ALS than are the axial muscle innervating medial motor column (MMC) or hypaxial motor column (HMC) motor neurons of the trunk. Whether this susceptibility is due to intrinsic properties of these motor neurons, their interactions with glia, or circuit and activity characteristics is not known.

One clear subtype-selective phenotype in ALS is the stereotypic order of degeneration of fast fatigable (FF), followed by fatigue-resistant (FR), and finally slow (S) motor units (Kanning et al. 2010). This sequence is evident in the morphological changes at the neuromuscular junctions of mutant SOD1 mice (Frey et al. 2000; Pun et al. 2006) and early loss of large diameter (FF) axons in ventral roots (Fischer et al. 2004). Loss of FF and FR motor units may be compensated for by sprouting from FR and S motor units, with resulting EMG and fiber type changes (Kanning et al. 2010). Data from human patients support this sequence of events: early signs of denervation in muscles, electromyograms consistent with denervation/reinnervation, and twitch force studies (Dengler et al. 1990; Fischer et al. 2004; de Carvalho et al. 2008). Despite these intriguing differences, the developmental mechanisms leading to FF, FR, or S motor unit motor neurons are not known. Furthermore, these motor neurons can be distinguished in situ by differences in cell soma size and axon caliber but are not unambiguously distinguished by any molecular or any genetic markers.

Two motor pools, however, show remarkable resilience even at end stages of disease. ALS patient autopsy samples show significant preservation of both oculomotor (Gizzi et al. 1992; Kaminski et al. 2002) and Onuf's nuclei (Mannen et al. 1977; Schroder and Reske-Nielsen 1984; Mannen 2000). Furthermore, these phenotypes translate robustly to ALS mice, where several eye muscle-innervating nuclei, including the oculomotor, are also preserved (Ferrucci et al. 2010). These strong, motor pool-specific ALS-resistance phenotypes suggest that, if human motor neurons could be induced to adopt an oculomotor phenotype in vitro, they should show resistance to disease-relevant stimuli and could therefore provide a strong validation of proposed models of ALS in the culture dish. However, there are currently no established protocols for generation of these midbrain motor neurons from ES or iPS cells.

Need for a Humanized Model of ALS

Despite extensive mechanistic knowledge about ALS, and over 30 Phase II and III clinical trials, there is only one marginally effective therapy for the disease (Hardiman et al. 2011). Riluzole, the only FDA-approved treatment for ALS, extends median survival by 2–3 months and is quite expensive (Miller et al. 2007). Riluzole is an anti-excitotoxic compound that is believed to act by decreasing the presynaptic release of glutamate onto motor neurons as well as persistent inward current in motor neurons themselves, thus decreasing overall excitation (Wang et al. 2004; Schuster et al. 2011). Other work suggests that riluzole may also be involved in increasing the production of neurotrophic factors in astrocytes, suggesting more than one potential mechanism of action (Peluffo et al. 1997; Mizuta et al. 2001). The high failure rate of drug trials in ALS is likely due to multiple factors, including initiation of treatment late in the course of disease, lack of biomarkers for diagnosis and treatment monitoring, heterogeneous patient populations, unknown etiology in many sporadic ALS patients, and pre-clinical development largely in non-human model systems (Gordon and Meininger 2011). Above all, it reflects the lack of validated therapeutic targets, i.e., molecular events whose inhibition can delay onset or slow progression.

Patient-specific iPS cell-derived motor neurons (iPS-MNs) potentially provide a humanized model of ALS that may help to address some of the above concerns. By virtue of replicating the exact genetic makeup of the donor patient, ALS iPS-MNs express endogenous levels of disease-causing genes and capture individual heterogeneity within disease. Thus, they may be better substrates for the identification or validation of therapeutic targets. It may also be possible to model individualized correlates of disease severity or sensitivity to particular environmental factors using iPS-MNs (e.g., organophosphate chemicals in the context of paraoxonase mutations; Ticozzi et al. 2010). Furthermore, since iPS-MNs may be considered to be at pre-symptomatic disease stages, they may be useful tools for the discovery of early causal events in ALS or biomarkers that are present before clinical symptoms. These applications could be critical for identifying both new therapeutic targets and patients for early initiation of treatment.

Although patient-specific iPS-MNs are a promising new model system for the study of ALS, the iPS technology is in its infancy. There are still many basic questions that need to be answered regarding the reliability of iPS cells as a whole before moving on to modeling disease using these cells. To place such concerns about iPS cells in context, we review below the initial characterization of iPS cells, the differentiation of iPS and other stem cells into motor neurons, and their strengths and weaknesses for modeling ALS.

Anatomy and Molecular Markers of Human Motor Neurons and Their Subtypes

To reliably model human motor neuron susceptibility in vitro, it is necessary to be certain that the cells that are generated conform to what is known about the timing of motor neuron generation in situ and about potential markers that can be used to identify human motor neurons in culture. Human motor neuron differentiation has mostly been followed by histological criteria (Rath et al. 1982; Altman and Bayer 2001; Bayer and Altman 2002). They are arranged in the same broad categories as in mouse, medial motor neurons innervating axial muscles (MMC, HMC), intermediolateral preganglionic motor neurons (PGC) and lateral column motor neurons innervating ventral and dorsal limb muscles (LMCm, LMCl). However, molecular tools and tracing studies have not been used to definitively establish a distinction between the MMC and HMC, the divisions of the LMC, or specific motor pools.

At the molecular level, in situ probes for the motor neuron marker HB9 identified motor neurons in the anterior horn of the spinal cord at Carnegie stage (CS) 15, corresponding to 35–38 days of development (Ross et al. 1998; Hagan et al. 2000). The authors also report that signal was seen at CS14 (31–35 days) but not at CS12 (26–30 days). These data validate the use of HB9 as a selective marker for embryonic human motor neurons and show that they appear at day 35–38 and perhaps as early as day 31–35. The ventricular zone HB9 signal was present only up to CS19 (days 48–51), suggesting the end of lumbar motor neurogenesis. The period of human motor neurogenesis is thus likely to occur in vivo over a period of almost 3 weeks, i.e., embryonic days 31 to 51.

Histological analysis offers another perspective on the period of motor neurogenesis, suggesting essentially the same start but perhaps an earlier conclusion. Incipient motor neurons were recognized by Ramon y Cajal in the human spinal cord at week 4 as early as 1909. More recently a detailed analysis of dozens of archival embryos was used to construct a comprehensive histological account of spinal development from gestational week (GW) 4 to gestational month 4 (Bayer and Altman 2002). These authors clearly identify motor neurons that have migrated out of the subventricular germinal zone by GW 4.5, and they interpret a subsequent thinning of the subventricular neuroepithelium – the presumptive human motor neuron progenitor domain – at GW 5.5 as indication that motor neurogenesis is largely complete. This particular estimate, however, derives from the cervical spinal cord only and is only a few days in advance of the loss of subventricular HB9 expression described above. In summary, the period of human motor neurogenesis cannot be definitively established with these tools alone, but the histological and molecular evidence supports the idea that human motor neurons are born between about embryonic day 30 and 50, corresponding well with the timing of the appearance of human motor neurons in classical hES cell differentiation protocols (see below).

Specification of Motor Neurons in Vitro from ES Cells

Procedures to derive motor neurons from mouse and hES cells recapitulate the major developmental pathways of motor neuron specification. Sequential application of retinoic acid (RA) and sonic hedgehog (SHH) to neuralized mouse ES cells results in cultures that contain approximately 40 % HB9[+] motor neurons at 7 days in vitro (DIV) (Wichterle et al. 2002). These mouse ES cell-derived motor neurons (mES-MNs) express other canonical motor neuron markers such as Islet1/2 and components of the acetylcholine synthesis pathway such as VAChT and CHAT. When transplanted into the developing chick spinal cord, mES-MNs settle in the ventral horn and exit primarily through the ventral root, projecting to both axial and limb musculature (Wichterle et al. 2002). Due to the presence of relatively high levels of RA during differentiation, mES-MNs derived under the RA/SHH protocol assume a primarily cervical MMC identity. Most motor neurons are Lhx3-positive and express either Hoxc6 or Hoxa5 (Wichterle et al. 2002; Peljto et al. 2010). Survival of mES-MNs in culture is dependent upon neurotrophic factor signaling in a manner similar to embryonic primary motor neurons. Furthermore, mES-MNs show electrophysiological properties consistent with embryonic motor neuron identity, including spontaneous action potential firing and excitatory responses to the neurotransmitters glutamate, GABA, and glycine (Miles et al. 2004). After 5 days in culture, mES-MNs show synaptic connectivity with other motor neurons as well as muscle cells grown in co-culture (Miles et al. 2004). Taken together, these data demonstrate that mES-MNs exhibit molecular and functional properties strikingly similar to cervical motor neurons in the mouse.

ES cell-derived motor neurons have also been successfully differentiated from hES cells using a derivative of the mouse protocol. Sequential exposure of un-patterned, neuralized hES cells to RA and SHH is used as an in mouse protocol, but over a period of approximately 32 days as opposed to 7 (Fig. 1; Li et al. 2005). Still, motor neurons produced from hES cells (hES-MNs) are remarkably similar to those derived from mouse ES cells. hES-MNs express the expected markers of motor neuron identity such as HB9, Islet 1/2, and ChAT and are capable of firing action potentials, as well as projecting axons out of chick spinal cord to muscle following transplantation (Li et al. 2005, 2008; Lee et al. 2007). Unexpectedly, some hES-MNs generated under these protocols express the brachial-thoracic marker Hoxc8 (Li et al. 2005). One potential explanation for this more caudal phenotype in hES cell cultures is that FGF-2 is used as a neuralizing agent before the addition of RA and may bias motor neurons towards more caudal fates, even in the presence of RA (Li et al. 2005). Finally, hES-MN differentiation protocols have recently been developed using enhanced neuralization procedures (dual SMAD inhibition) that both shorten differentiation times (to 21 days from 32) and result in increased motor neuron differentiation efficiency (Chambers et al. 2009; Boulting et al. 2011).

Emerging protocols demonstrate that ES-MNs can also be prospectively patterned along the rostral-caudal axis to produce different classes of limb-innervating motor neurons. As mentioned above, the majority of mES-MNs generated under

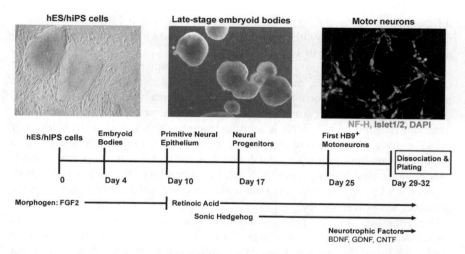

Fig. 1 Human stem cell differentiation to motor neurons. hESCs/iPSCs are differentiated to motor neurons through exposure to fibroblast growth factor-2 (FGF2), retinoic acid, and sonic hedgehog (or agonist) at developmentally appropriate time points. The neurotrophic factors brain-derived neurotrophic factor (BDNF), glia-derived neurotrophic factor (GDNF), and ciliary neurotrophic factor (CNTF) are added to increase motor neuron survival and maturation following differentiation. Following differentiation, iPS/hES motor neurons express the neuronal marker NF-H and the motor neuron marker Islet1

standard RA protocols are of cervical identity. Under the appropriate morphogen-free conditions, however, mES cells spontaneously differentiate into motor neurons of mostly brachial and thoracic identity, as evidenced by increased expression of Hoxc8 and Hoxc9 and decreased expression of Hoxa5 in HB9+ cells as compared to standard RA protocols (Peljto et al. 2010). Caudal motor neuron fates in these cultures are dependent upon endogenous SHH, FGF, and Wnt signaling and can be blocked by antagonists to these pathways. Furthermore, the addition of exogenous FGF-8 or GDF-11 to low RA motor neuron differentiations induces significant expression of the caudal Hox genes, Hoxc9 and Hoxd10, in motor neurons, respectively, demonstrating that mES-MNs can be patterned into specific rostrocaudal sub-types (Peljto et al. 2010).

Yield and Purity of hES Cell-Derived Motor Neurons

Scalability is a particularly useful feature of ES cell-derived motor neuron differentiation protocols. The average adult human has approximately 50,000 motor neurons in the spinal cord (Tomlinson and Irving 1977). Thus, even if primary cultures from post-mortem spinal cords were technically feasible, the low number of cells would be a limiting factor for many avenues of research. Fortunately, hES cell-derived motor neurons can be produced in much greater quantity. An average differentiation can yield up to 40 million cells, of which 20–35 % are motor neurons. Thus, one 21-day hES cell differentiation can yield motor neurons equivalent to the sum of those in 160–280 individuals.

Although motor neuron yield can be scaled to suit nearly any experimental design, low motor neuron purity in culture can be a confounding variable. The proportion of motor neurons in hES cell-derived cultures ranges from 10 % to 40 % of total cells when using current versions of motor neuron differentiation protocols. Thus, purification of hES-MNs is a critical step towards isolating populations of cells for study. Unfortunately, the p75 receptor, which is used to efficiently isolate motor neurons from primary rodent cultures, is not motor neuron-specific in hES-MN cultures. Additionally, since hES cells are refractory to genetic modification, expression of a sortable surface marker under the control of a motor neuron-specific promoter has proven difficult.

Nevertheless, motor neuron reporter strategies have been useful in circumventing some of the obstacles posed by low motor neuron yield. Two published hES-MN reporter lines exist, both expressing GFP from the motor neuron-specific HB9 promoter (Di Giorgio et al. 2008; Placantonakis et al. 2009). Another approach, utilizing a shortened version of the HB9 promoter, has also been implemented to label hES-MNs in culture. Studies of the HB9 promoter indicate that a great deal of its motor neuron-specific activity is contained in a 3.5 kb segment that is 5.5 kb upstream of the transcriptional start site (Lee et al. 2004; Nakano et al. 2005). Pairing this 3.5 kb segment with generic minimal promoters has proven a useful strategy for labeling hES-MNs in vitro by either plasmid transfection or lentiviral transduction (Singh Roy et al. 2005; Marchetto et al. 2008). These examples demonstrate the critical importance of genetic reporter approaches for identifying and studying human motor neurons and emphasize the need for further work to label motor neuron subpopulations and patient-derived hiPS-MNs.

Application of Stem Cell-Derived Motor Neurons to the Study of ALS

To study ALS in vitro, it is critical to compare motor neurons with and without the genetic determinants of disease. Several means to this end have been tested. Both genetic manipulation of hES cells and transient transfection of hES-MNs could be used to introduce ALS-causing genes into these cells and study their downstream effects. Recently developed zinc finger nuclease approaches will likely allow the directed mutagenesis of ALS-causing genes in hES cells (Urnov et al. 2010). Alternatively, transient transfection of hES-MNs with familial ALS-causing genes can also be achieved using current technology, albeit with poor control of transgene expression level (Karumbayaram et al. 2009). However, both of these approaches forfeit access to patient-specific genetic background, which has a known modifying effect in ALS, and to all sporadic forms of the disease.

The Yamanaka group first opened the door to an alternative strategy by demonstrating the feasibility of inducing pluripotency in somatic cells using a defined set of transcription factors. They retrovirally expressed a group of 24 master-regulators of ES cell fate in mouse fibroblasts and monitored these cells for re-activation of other known ES cell markers and morphological characteristics. Remarkably, multiple ES cell-like colonies appeared approximately 16 days after transduction (Takahashi and Yamanaka 2006). A simple subtractive analysis performed on the initial group of 24 genes yielded a set of 4 key factors for induced pluripotency: Oct-3/4, Sox-2, Klf-4, and c-Myc (Takahashi and Yamanaka 2006). Pluripotent cell lines derived by this method are dubbed induced pluripotent stem cells (iPS cells). iPS cells, although derived from fibroblasts or other somatic cells, express protein markers and DNA methylation patterns characteristic of ES cells. Furthermore, the first iPS cells passed both the teratoma and chimera tests of pluripotency and could be differentiated in vitro into derivatives of all three germ layers (Takahashi and Yamanaka 2006). Subsequent experiments confirmed the pluripotency of mouse iPS cells by producing tetraploid embryos derived entirely from these cells (Boland et al. 2009; Zhao et al. 2009).

Human cells can also be reprogrammed to pluripotent iPS cells (hiPS cells). Reprogramming of dermal fibroblasts has been reported with the both the Yamanaka and Thomson four-factor cocktails, Oct-3/4, KLF-2 and Sox-2 without c-MYC, as well as Oct-3/4 and Sox-2 alone in the presence of valproic acid (Takahashi et al. 2007; Yu et al. 2007; Huangfu et al. 2008; Park et al. 2008). Reprogramming of human dermal fibroblasts occurs over a slightly longer time period than in mouse, likely due to slower cell division rates in hES cells and fibroblasts.

hiPS cells have generated enormous excitement due to their potential applications in disease modeling and regenerative therapy. Directed differentiation of stem cells provides the first access to living preparations of many patient-specific tissue types. By capturing patient-specific genetic backgrounds, iPS cells will enable modeling of poorly understood complex genetic disorders, creation of humanized disease models that may be used to study disease modifiers and other correlations with clinical data, and immunologically matched tissue for eventual cell-replacement therapy. These advantages of patient-derived iPS cells are particularly promising for ALS research (Fig. 2). Using established techniques, iPS cells derived from ALS patients were shown to differentiate into motor neurons (ALS iPS-MNs; Dimos et al. 2008). Importantly, ALS iPS-MNs provide the only access to pre-symptomatic human ALS motor neurons. These cells express endogenous levels of ALS-causing genes in the context of an individual-specific genetic background. However, before investing fully in them as a tool for intensive study, it is important to address their potential drawbacks.

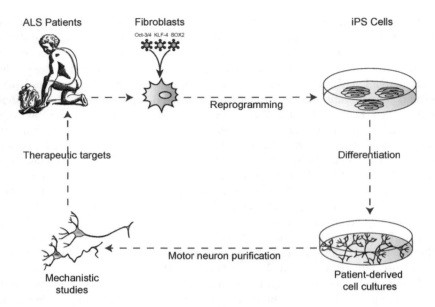

Fig. 2 Building an iPS cell-based model of ALS. An overview of how iPS cells may be used to model ALS. Fibroblasts (or other donor tissues) are derived from ALS patients and reprogrammed into iPS cells. iPS cells are then differentiated into mixed cultures of motor neurons and other cells. Next, motor neurons would ideally be isolated from these cultures before use in disease modeling experiments. Insight gained into disease mechanisms through such studies may point towards new therapeutic targets. ALS Patients image from Descartes (1664)

Concerns About Using iPS Cells as an Approach for Modeling ALS

Before assessing ALS-related phenotypes in iPS-MNs, it is necessary to more robustly establish the iPS system as a reliable alternative to the gold standard of hES cells. Some concerns pertain to retention of epigenetic marks during reprogramming, transcriptional differences between iPS and ES cells, potential deleterious effects of sustained transgene expression, accumulation of somatic mutations, and reduced efficiency of directed differentiation.

Several studies have indicated that many iPS lines are incompletely reprogrammed and retain a memory of the cell type of origin. During iPS cell derivation, nascent iPS lines pass through a partially reprogrammed state before becoming fully pluripotent. Some iPS lines may become trapped in this state and never attain true pluripotency (Mikkelsen et al. 2008). Those lines that do become fully pluripotent can retain epigenetic marks of the tissue of origin and exhibit reduced differentiation efficiency towards other cell lineages, especially at early passages (Kim et al. 2010). Furthermore, parentally imprinted genes may never be properly re-programmed during iPS cell derivation (Stadtfeld et al. 2010). However, with continued passaging, the transcriptional state of both human and mouse iPS cells begins to converge upon that of ES cells, although there are still some iPS-specific differences in expression (Chin et al. 2009). It has been suggested that expression of retrovirally integrated reprogramming genes is responsible for some of the differences in global gene expression between iPS and ES cells. In support of this idea, the overall transcriptional similarity between hES and hiPS cells significantly increases following LoxP-mediated excision of reprogramming factors (Soldner et al. 2009).

Perhaps the most important concern for ALS modeling is that it has been suggested that iPS cells may show a specific deficit in forming neural tissues and motor neurons. One study differentiated a set of 12 iPS cell lines and 5 hES cell lines to neural epithelia and observed that 11 of 12 iPS cell lines formed Pax6-positive neural progenitors at reduced efficiency compared to ES cell lines (Hu et al. 2010). Furthermore, a subset of these iPS cell lines also showed a diminished capacity to form HB9$^+$ motor neurons as compared to a single hES cell line. However, these same iPS cell lines formed Olig2$^+$ motor neuron progenitors more efficiently than the hES cell line. The reason for this discrepancy is unclear, but the results suggest that iPS cells may be a difficult starting point from which to make neurons. One potential confounding variable in these experiments, however, is that the hES cell line used in the production of motor neurons (H9) has a particularly high neurogenic potential (Osafune et al. 2008; Bock et al. 2011). Furthermore, it has recently been shown that longer times in culture seem to increase the efficiency of neuralization in iPS cells, suggesting that the reduced efficiencies observed by Hu et al. (2010) may be a transient effect (Koehler et al. 2011).

To address these questions, we collaborated with Dr. Kevin Eggan's lab (Harvard University) to perform an extensive phenotypic characterization of hiPS cells in direct comparison to hES cells before initiating disease-modeling research. To accomplish this, we generated a panel of iPS lines from fibroblasts donated by ALS patients and healthy controls. We then performed a broad characterization of their pluripotency and differentiation capacity, comparing results from independent experiments in two different laboratories (Boulting et al. 2011). In this study, we evaluated pluripotency and efficiency of directed differentiation to motor neurons in a test set of 5 hES and 16 hiPS cell lines (Boulting et al. 2011). In addition to comparing ES and iPS cell lines, the test-set was designed to assess variability in iPSC lines conferred by individual donor identity, iPS cell clone, 3 versus 4-factor reprogramming, age, sex, and ALS genotype. We demonstrated that one in five iPS cell lines had a serious defect in neural differentiation, but these were easily detectable and could be corrected by early neuralization using dual SMAD inhibition. The remaining iPS cell lines generated motor neurons with positional identities and functional electrophysiological behavior equivalent to ES cell-derived motor neurons. Finally, we excluded several of the demographic variables – age, disease status, number of reprogramming genes – as sources of variation among iPS-MNs but identified donor genetic background (including and potentially limited to sex) as a significant factor.

These studies provide a resource of disease and control cell lines for studying ALS and also provide practical guidelines for the derivation and analysis of iPS cell lines. Most importantly, they provide confidence that iPS cell lines can generate differentiated motor neurons with comparable functional characteristics to those derived from ES cell lines and set the stage for making models of genetic disease. The performance of individual iPS cell lines was remarkably similar between the two different labs involved in our study, which shows that lines retain intrinsic properties across passages, and their performance in well-defined differentiation protocols (with at least four different operators) was highly robust. This finding gives confidence that hIPS cells are reproducible tools, an important criterion required for cellular substrates of disease modeling and drug screening.

A companion study addressed variability in iPS cell differentiation efficiency using a genome-wide profiling strategy (Bock et al. 2011). By measuring the expression of a series of lineage markers during morphogen-free random differentiation, the Meissner group was able to compute a scorecard of lineage differentiation potential for a panel of iPS lines. Our iPS-MN differentiation efficiency data were then used as one means of validating this scorecard. A comparison of the technical requirements and limitations of both approaches can match their strengths to the needs of a specific research program. To use either assay, operators will first need to be comfortable with basic hES/iPS cell culture techniques as well as any protocols necessary to make a particular differentiated derivative of interest. The advantages of the scorecard approach are (1) limited operator skill required to perform random differentiation; (2) savings of time and reagent costs during differentiation stages (16 days in Bock et al. 2011 as opposed to 21–32 days in Boulting et al. 2011); (3) measurement of differentiation potential to all three germ layers; and (4) much greater scalability than directed differentiation. In contrast, the advantages of the directed differentiation approach are (1) direct measurement of differentiation efficiency for the particular cell type of interest; (2) no need to normalize and interpret data from scorecard assays or to develop cutoff values to choose between lines; (3) no need to buy scorecard chip reagents; and (4) it utilizes only existing techniques and reagents. Thus, for use in a specialist lab concerned only with the production of one cell type or germ layer, directed differentiation to the particular cell type being studied seems the most straightforward and cost-effective method. However, for core facilities, groups working with large numbers of iPS cell lines, or groups deriving tissues belonging to multiple germ layers, the scorecard assay offers greater scalability and simultaneous information about differentiation efficiency to all three germ layers.

Conclusion and Perspectives

We have seen that iPS cells can be generated from mouse and human somatic cells and that these can reproduce the cardinal functional characteristics of ES cells. Importantly, they can generate the cell types affected in many diseases and have already been used to generate highly relevant new model systems for a variety of developmental disorders and diseases. Several challenges and questions remain from a disease modeling perspective, however. Most generally, do human iPS cells regain the tabula rasa of ES that will allow them to navigate in vitro the normal course of development to generate extremely specific differentiated derivatives? Specifically, in the context of ALS and motor neurons, will iPS cells demonstrate responsiveness to multiple patterning cues to enact programs of diversification into dozens of specific and coherent motor neuron subtypes? Finally, will iPS cell-derived neurons, with presumably an embryonic age, be able to mature sufficiently to reproduce disease-relevant phenotypes that take decades to develop in vivo? Work aimed at answering these questions is currently underway in our laboratory as well as others and may have bearing not only on how ALS may be modeled using this new resource but also in the ways age-related neurodegenerative disorders can be studied through the lens of iPS cells.

Acknowledgments Work in the authors' laboratory was supported by Project A.L.S., P2ALS, The Tow Foundation, the SMA Foundation, NYSTEM, and NINDS. We thank our colleagues in the Project A.L.S., Eggan, Wichterle and Henderson laboratories for many stimulating discussions.

References

Altman J, Bayer SA (2001) Development of the human spinal cord: an interpretation based on experimental studies in animals. Oxford University Press, Oxford

Bayer SA, Altman J (2002) The spinal cord from gestational week 4 to the 4th postnatal month. Taylor and Francis, Boca Raton

Bock C, Kiskinis E, Verstappen G, Gu H, Boulting G, Smith ZD, Ziller M, Croft GF, Amoroso MW, Oakley DH, Gnirke A, Eggan K, Meissner A (2011) Reference maps of human ES and iPS cell variation enable high-throughput characterization of pluripotent cell lines. Cell 144:439–452

Boillée S, Vande Velde C, Cleveland DW (2006) ALS: a disease of motor neurons and their nonneuronal neighbors. Neuron 52:39–59

Boland MJ, Hazen JL, Nazor KL, Rodriguez AR, Gifford W, Martin G, Kupriyanov S, Baldwin KK (2009) Adult mice generated from induced pluripotent stem cells. Nature 461:91–94

Boulting GL, Kiskinis E, Croft GF, Amoroso MW, Oakley DH, Wainger BJ, Williams DJ, Kahler DJ, Yamaki M, Davidow L, Rodolfa CT, Dimos JT, Mikkilineni S, MacDermott AB, Woolf CJ, Henderson CE, Wichterle H, Eggan K (2011) A functionally characterized test set of human induced pluripotent stem cells. Nature Biotechnol 29:279–286

Chambers SM, Fasano CA, Papapetrou EP, Tomishima M, Sadelain M, Studer L (2009) Highly efficient neural conversion of human ES and iPS cells by dual inhibition of SMAD signaling. Nature Biotechnol 27:275–280

Charcot J-M, Joffroy A (1869) Deux cas d'atrophie musculaire progressive: avec lésions de la substance grise et des faisceaux antéro-latéraux de la moelle épinière. V. Masson, Paris

Chin MH, Mason MJ, Xie W, Volinia S, Singer M, Peterson C, Ambartsumyan G, Aimiuwu O, Richter L, Zhang J, Khvorostov I, Ott V, Grunstein M, Lavon N, Benvenisty N, Croce CM, Clark AT, Baxter T, Pyle AD, Teitell MA, Pelegrini M, Plath K, Lowry WE (2009) Induced pluripotent stem cells and embryonic stem cells are distinguished by gene expression signatures. Cell Stem Cell 5:111–123

Clement AM, Nguyen MD, Roberts EA, Garcia ML, Boillée S, Rule M, McMahon AP, Doucette W, Siwek D, Ferrante RJ, Brown RH Jr, Julien JP, Goldstein LS, Cleveland DW (2003) Wild-type nonneuronal cells extend survival of SOD1 mutant motor neurons in ALS mice. Science 302:113–117

de Carvalho MA, Pinto S, Swash M (2008) Paraspinal and limb motor neuron involvement within homologous spinal segments in ALS. Clin Neurophysiol 119:1607–1613

Dejesus-Hernandez M, Mackenzie IR, Boeve BF, Boxer AL, Baker M, Rutherford NJ, Nicholson AM, Finch NA, Flynn H, Adamson J, Kouri N, Wojtas A, Sengdy P, Hsiung GY, Karydas A, Seeley WW, Josephs KA, Coppola G, Geschwind DH, Wszolek ZK, Feldman H, Knopman DS, Petersen RC, Miller BL, Dickson DW, Boylan KB, Graff-Radford NR, Rademakers R (2011) Expanded GGGGCC hexanucleotide repeat in noncoding region of C9ORF72 causes chromosome 9p-linked FTD and ALS. Neuron 72:245–256

Dengler R, Konstanzer A, Küther G, Hesse S, Wolf W, Struppler A (1990) Amyotrophic lateral sclerosis: macro-EMG and lateral sclerosis: macro-EMG and twitch forces of single motor units. Muscle Nerve 13:545–550

Descartes R (1664) L'Homme. Charles Angot, Paris

Di Giorgio FP, Boulting GL, Bobrowicz S, Eggan KC (2008) Human embryonic stem cell-derived motor neurons are sensitive to the toxic effect of glial cells carrying an ALS-causing mutation. Cell Stem Cell 3:637–648

Dimos JT, Rodolfa KT, Niakan KK, Weisenthal LM, Mitsumoto H, Chung W, CroftGF SG, Leibel R, Goland R, Wichterle H, Henderson CE, Eggan K (2008) Induced pluripotent stem cells generated from patients with ALS can be differentiated into motor neurons. Science 321:1218–1221

Ferrucci M, Spalloni A, Bartalucci A, Cantafora E, Fulceri F, Nutini M, Longone P, Paparelli A, Fornai F (2010) A systematic study of brainstem motor nuclei in a mouse model of ALS, the effects of lithium. Neurobiol Disease 37:370–383

Fischer LR, Culver DG, Tennant P, Davis AA, Wang M, Castellano-Sanchez A, Khan J, Polak MA, Glass JD (2004) Amyotrophic lateral sclerosis is a distal axonopathy: evidence in mice and man. Exp Neurol 185:232–240

Frey D, Schneider C, Xu L, Borg J, Spooren W, Caroni P (2000) Early and selective loss of neuromuscular synapse subtypes with low sprouting competence in motoneuron diseases. J Neurosci 20:2534–2542

Gitcho MA, Baloh RH, Chakraverty S, Mayo K, Norton JB, Levitch D, Hatanpaa KJ, White CL 3rd, Bigio EH, Caselli R, Baker M, Al-Lozi MT, Morris JC, Pestronk A, Rademakers R, Goate AM, Cairns NJ (2008) TDP-43 A315T mutation in familial motor neuron disease. Ann Neurol 63:535–538

Gizzi M, DiRocco SM, Cohen B (1992) Ocular motor function in motor neuron disease. Neurology 42:1037–1046

Gordon PH, Meininger V (2011) How can we improve clinical trials in amyotrophic lateral sclerosis? Nature Rev Neurol 7:650–654

Hagan DM, Ross AJ, Strachan T, Lynch SA, Ruiz-Perez V, Wang YM, Scambler P, Custard E, Reardon W, Hassan S, Nixon P, Papapetrou C, Winter RM, Edwards Y, Morrison K, Barrow M, Cordier-Alex MP, Correia P, Galvin-Parton PA, Gaskill S, Gaskin KJ, Garcia-Minaur S, Gereige R, Hayward R, Homfray T (2000) Mutation analysis and embryonic expression of the HLXB9 Currarino syndrome gene. Am J Hum Genet 66:1504–1515

Hardiman O, van den Berg LH, Kiernan MC (2011) Clinical diagnosis and management of amyotrophic lateral sclerosis. Nature Rev Neurol 7:639–649

Hu B-Y, Weick JP, Yu J, Ma LX, Zhang XQ, Thomson JA, Zhang SC (2010) Neural differentiation of human induced pluripotent stem cells follows developmental principles but with variable potency. Proc Natl Acad Sci USA 107:4335–4340

Huangfu D, Osafune K, Maehr R, Guo W, Eijkelenboom A, Chen S, Muhlestein W, Melton DA (2008) Induction of pluripotent stem cells from primary human fibroblasts with only Oct4 and Sox2. Nature Biotechnol 26:1269–1275

Kabashi E, Valdmanis PN, Dion P, Spiegelman D, McConkey BJ, Vande Velde C, Bouchard JP, Lacomblez L, Pochigaeva K, Salachas F, Pradat PF, Camu W, Meininger V, Dupre N, Rouleau GA (2008) TARDBP mutations in individuals with sporadic and familial amyotrophic lateral sclerosis. Nature Genet 40:572–574

Kaminski HJ, Richmonds CR, Kusner LL, Mitsumoto H (2002) Differential susceptibility of the ocular motor system to disease. Ann NY Acad Sci 956:42–54

Kanning KC, Kaplan A, Henderson CE (2010) Motor neuron diversity in development and disease. Ann Rev Neurosci 33:409–440

Karumbayaram S, Kelly TK, Paucar AA, Roe AJ, Umbach JA, Charles A, Goldman SA, Kornblum HI, Wiedau-Pazos M (2009) Human embryonic stem cell-derived motor neurons expressing SOD1 mutants exhibit typical signs of motor neuron degeneration linked to ALS. Dis Model Mech 2:189–195

Kim K, Doi A, Wen B, Ng K, Zhao R, Cahan P, Kim J, Aryee MJ, Ji H, Ehrlich LI, Yabuuchi A, Takeuchi A, Cunniff KC, Hongguang H, McKinney-Freeman S, Naveiras O, Yoon TJ, Irizarry RA, Jung N, Seita J, Hanna J, Murakami P, Jaenisch R, Weissleder R, Orkin SH, Weissman IL, Feinberg AP, Daley GQ (2010) Epigenetic memory in induced pluripotent stem cells. Nature 467:285–290

Koehler KR, Tropel P, Theile JW, Kondo T, Cummins TR, Viville S, Hashino E (2011) Extended passaging increases the efficiency of neural differentiation from induced pluripotent stem cells. BMC Neuro 12:82

Lee SK, Jurata LW, Funahashi J, Ruiz EC, Pfaff SL (2004) Analysis of embryonic motoneuron gene regulation: derepression of general activators function in concert with enhancer factors. Development 131:3295–3306

Lee H, Shamy GA, Elkabetz Y, Schofield CM, Harrsion NL, Panagiotakos G, Socci ND, Tabar V, Studer L (2007) Directed differentiation and transplantation of human embryonic stem cell-derived motoneurons. Stem Cells 25:1931–1939

Li XJ, Du ZW, Zarnowska ED, Pankratz M, Hansen LO, Pearce RA, Zhang SC (2005) Specification of motoneurons from human embryonic stem cells. Nature Biotechnol 23:215–221

Li XJ, Hu BY, Jones SA, Zhang YS, Lavaute T, Du ZW, Zhang SC (2008) Directed differentiation of ventral spinal progenitors and motor neurons from human embryonic stem cells by small molecules. Stem Cells 26:886–893

Mannen T (2000) Neuropathological findings of Onuf's nucleus and its significance. Neuropathology 20(Suppl):S30–S33

Mannen T, Iwata M, Toyokura Y, Nagashima K (1977) Preservation of a certain motoneurone group of the sacral cord in amyotrophic lateral sclerosis: its clinical significance. J Neurol Neurosurg Psychiat 40:464–469

Marchetto MC, Muotri AR, Mu Y, Smith AM, Cezar GG, Gage FH (2008) Non-cell-autonomous effect of human SOD1 G37R astrocytes on motor neurons derived from human embryonic stem cells. Cell Stem Cell 3:649–657

Mikkelsen TS, Hanna J, Zhang X, Ku M, Wernig M, Schorderet P, Bernstein BE, Jaenisch R, Lander ES, Meissner A (2008) Dissecting direct reprogramming through integrative genomic analysis. Nature 454:49–55

Miles GB, Yohn DC, Wichterle H, Jessell TM, Rafuse VF, Brownstone RM (2004) Functional properties of motoneurons derived from mouse embryonic stem cells. J Neurosci 24:7848–7858

Miller RG, Mitchell JD, Moore DH (2007) Riluzole for amyotrophic lateral sclerosis (ALS)/motor neuron disease (MND). Cochrane Database Syst Rev 1:CD001447

Mizuta I, Ohta M, Ohta K, Nishimura M, Mizuta E, Kuno S (2001) Riluzole stimulates nerve growth factor, brain-derived neurotrophic factor and glial cell line-derived neurotrophic factor synthesis in cultured mouse astrocytes. Neurosci Lett 310:117–120

Nakano T, Windrem M, Zappavigna V, Goldman SA (2005) Identification of a conserved 125 base-pair Hb9 enhancer that specifies gene expression to spinal motor neurons. Dev Biol 283:474–485

Osafune K, Caron L, Borowiak M, Martinez RJ, Fitz-Gerald CS, Sato Y, Cowan CA, Chien KR, Melton DA (2008) Marked differences in differentiation propensity among human embryonic stem cell lines. Nature Biotechnol 26:313–315

Park I-H, Zhao R, West JA, Yabuuchi A, Huo H, Ince TA, Lerou PH, Lensch MW, Daley GQ (2008) Reprogramming of human somatic cells to pluripotency with defined factors. Nature 451:141–146

Peljto M, Dasen JS, Mazzoni EO, Jessell TM, Wichterle H (2010) Functional diversity of ESC-derived motor neuron subtypes revealed through intraspinal transplantation. Cell Stem Cell 7:355–366

Peluffo H, Estevez A, Barbeito L, Stutzmann JM (1997) Riluzole promotes survival of rat motoneurons in vitro by stimulating trophic activity produced by spinal astrocyte monolayers. Neurosci Lett 228:207–211

Placantonakis DG, Tomishima MJ, Lafaille F, Desbordes SC, Jia F, Socci ND, Viale A, Lee H, Harrison N, Tabar V, Studer L (2009) BAC transgenesis in human embryonic stem cells as a novel tool to define the human neural lineage. Stem Cells 27:521–532

Pun S, Santos AF, Saxena S, Xu L, Caroni P (2006) Selective vulnerability and pruning of phasic motoneuron axons in motoneuron disease alleviated by CNTF. Nature Neurosci 9:408–419

Rath G, Gopinath G, Bijlani V (1982) Prenatal development of the human spinal cord I. Ventral motor neurons. J Neurosci Res 7:437–441

Ravits JM, La Spada R (2009) ALS motor phenotype heterogeneity, focality, and spread: deconstructing motor neuron degeneration. Neurology 73:805–811

Ravits J, Paul P, Jorg C (2007) Focality of upper and lower motor neuron degeneration at the clinical onset of ALS. Neurology 68:1571–1575

Renton AE, Majounie E, Waite A, Simón-Sánchez J, Rollinson S, Gibbs JR, Schymick JC, Laaksovirta H, van Swieten JC, Myllykangas L, Kalimo H, Paetau A, Abramzon Y, Remes AM, Kaganovich A, Scholz SW, Duckworth J, Ding J, Harmer DW, Hernandez DG, Johnson JO, Mok K, Ryten M, Trabzuni D, Guerreiro RJ, Orrell RW, Neal J, Murray A, Pearson J, Jansen IE, Sondervan D, Seelaar H, Blake D, Young K, Halliwell N, Callister JB, Toulson G, Richardson A, Gerhard A, Snowden J, Mann D, Neary D, Nalls MA, Peuralinna T, Jansson L, Isoviita VM, Kaivorinne AL, Hölttä-Vuori M, Ikonen E, Sulkava R, Benatar M, Wuu J, Chiò A, Restagno G, Borghero G, Sabatelli M; ITALSGEN Consortium, Heckerman D, Rogaeva E, Zinman L, Rothstein JD, Sendtner M, Drepper C, Eichler EE, Alkan C, Abdullaev Z, Pack SD, Dutra A, Pak E, Hardy J, Singleton A, Williams NM, Heutink P, Pickering-Brown S, Morris HR, Tienari PJ, Traynor BJ (2011) A hexanucleotide repeat expansion in C9ORF72 Is the cause of chromosome 9p21-linked ALS-FTD. Neuron 72:257–268

Rosen DR, Siddique T, Patterson D, Figlewicz DA, Sapp P, Hentati A, Donaldson D, Goto J, O'Regan JP, Deng HX, Rahmani Z, Krizus A, McKenna-Yasek D, Cayabyab A, Gaston SM, Berger R, Tanzi RE, Halperin JJ, Herzfeldt B, Van den Bergh R, Hung WY, Bird B, Deng G, Mulder DW, Smyth C, Laing NG, Soriano E, Pericak-Vance MA, Haines J, Rouleau GA, Gusella JS, Horvitz RH, Brown RH Jr (1993) Mutations in Cu/Zn superoxide dismutase gene are associated with familial amyotrophic lateral sclerosis. Nature 362:59–62

Ross AJ, Ruiz-Perez V, Wang Y, Hagan DM, Scherer S, Lynch SA, Lindsay S, Custard E, Belloni E, Wilson DI, Wadey R, Goodman F, Orstavik KH, Monclair T, Robson S, Reardon W, Burn J, Scambler P, Strachan T (1998) A homeobox gene, HLXB9, is the major locus for dominantly inherited sacral agenesis. Nature Genet 20:358–361

Schroder HD, Reske-Nielsen E (1984) Preservation of the nucleus X-pelvic floor motosystem in amyotrophic lateral sclerosis. Clin Neuropathol 3:210–216

Schuster JE, Fu R, Siddique T, Heckman CJ (2011) The effect of prolonged Riluzole exposure on cultured motoneurons in a mouse model of ALS. J Neurophysiol 107:484–492

Singh Roy N, Nakano T, Xuing L, Kang J, Nedergaard M, Goldman SA (2005) Enhancer-specified GFP-based FACS purification of human spinal motor neurons from embryonic stem cells. Exp Neurol 196:224–234

Soldner F, Hockemeyer D, Beard C, Gao Q, Bell GW, Cook EG, Hargus G, Blak A, Cooper O, Mitalipova M, Isacson O, Jaenisch R (2009) Parkinson's disease patient-derived induced pluripotent stem cells free of viral reprogramming factors. Cell 136:964–977

Stadtfeld M, Apostolou E, Akutsu H, Fukuda A, Follett P, Natesan S, Kono T, Shioda T, Hochedlinger K (2010) Aberrant silencing of imprinted genes on chromosome 12qF1 in mouse induced pluripotent stem cells. Nature 465:175–181

Takahashi K, Yamanaka S (2006) Induction of pluripotent stem cells from mouse embryonic and adult fibroblast cultures by defined factors. Cell 126:663–676

Takahashi K, Tanabe K, Ohnuki M, Narita M, Ichisaka T, Tomoda K, Yamanaka S (2007) Induction of pluripotent stem cells from adult human fibroblasts by defined factors. Cell 131:861–872

Ticozzi N, Leclerc AL, Keagle PJ, Glass JD, Wills AM, van Blitterswijk M, Bosco DA, Rodriguez-Leyva I, Gellera C, Ratti A, Taroni F, McKenna-Yasek D, Sapp PC, Silani V, Furlong CE, Brown RH Jr, Landers JE (2010) Paraoxonase gene mutations in amyotrophic lateral sclerosis. Ann Neurol 68:102–107

Tomlinson BE, Irving D (1977) The numbers of limb motor neurons in the human lumbosacral cord throughout life. J Neurol Sci 34:213–219

Urnov FD, Rebar EJ, Holmes MC, Zhang HS, Gregory PD (2010) Genome editing with engineered zinc finger nucleases. Nature Rev Genet 11:636–646

Van Deerlin VM, Leverenz JB, Bekris LM, Bird TD, Yuan W, Elman LB, Clay D, Wood EM, Chen-Plotkin AS, Martinez-Lage M, Steinbart E, McCluskey L, Grossman M, Neumann M, Wu IL, Yang WS, Kalb R, Galasko DR, Montine TJ, Trojanowski JQ, Lee VM, Schellenberg GD, Yu CE (2008) TARDBP mutations in amyotrophic lateral sclerosis with TDP-43 neuropathology: a genetic and histopathological analysis. Lancet Neurol 7:409–416

van Es MA, Veldink JS, Saris CG, Blauw HM, van Vught PW, Birve A, Lemmens R, Schelhaas HJ, Groen EJ, Huisman MH, van der Kooi AJ, de Visser M, Dahlberg C, Estrada K, Rivadeneira F, Hofman A, Zwarts MJ, van Doormaal PT, Rujescu D, Strengman E, Giegling I, Muglia P, Tomik B, Slowik A, Uitterlinden AG, Hendrich C, Waibel S, Meyer T, Ludolph AC, Glass JD, Purcell S, Cichon S, Nöthen MM, Wichmann HE, Schreiber S, Vermeulen SH, Kiemeney LA, Wokke JH, Cronin S, McLaughlin RL, Hardiman O, Fumoto K, Pasterkamp RJ, Meininger V, Melki J, Leigh PN, Shaw CE, Landers JE, Al-Chalabi A, Brown RH Jr, Robberecht W, Andersen PM, Ophoff RA, van den Berg LH (2009) Genome-wide association study identifies 19p13.3 (UNC13A) and 9p21.2 as susceptibility loci for sporadic amyotrophic lateral sclerosis. Nature Genet 41:1083–1087

Wang SJ, Wang KW, Wang WC (2004) Mechanisms underlying the riluzole inhibition of glutamate release from rat cerebral cortex nerve terminals (synaptosomes). Neuroscience 125:191–201

Wichterle H, Lieberam I, Porter JA, Jessell TM (2002) Directed differentiation of embryonic stem cells into motor neurons. Cell 110:385–397

Yu J, Vodyanik MA, Smuga-Otto K, Antosiewicz-Bourget J, Frane JL, Tian S, Nie J, Jonsdottir GA, Ruotti V, Stewart R, Slukvin II, Thomson JA (2007) Induced pluripotent stem cell lines derived from human somatic cells. Science 318:1917–1920

Zhao XY, Li W, Lv Z, Liu L, Tong M, Hai T, Hao J, Guo CL, Ma QW, Wang L, Zeng F, Zhou Q (2009) iPS cells produce viable mice through tetraploid complementation. Nature 461:86–90

Using Pluripotent Stem Cells to Decipher Mechanisms and Identify Treatments for Diseases That Affect the Brain

Marc Peschanski and Cécile Martinat

Abstract Pluripotent stem cell lines derived from donors who carry a mutant gene at the origin of a monogenic disease can be obtained currently either from embryos that are characterized as gene carriers during a pre-implantation genetic diagnosis (PGD) procedure or through genetic reprogramming of donors' sample cells. Both methods can be used to screen libraries of compounds in a search for new treatments. Currently, the embryonic stem (ES) cell bank at I-Stem comprises over 30 PGD-derived cell lines representing over 15 diseases. Induced pluripotent stem (iPS) cell lines are derived upon request and I-Stem welcomes colleagues who need iPS cell lines in its "iPS workshop" (see www.istem.eu). Robust read-outs relevant to the pathological mechanisms should first be identified. On this basis, a screening platform either in high throughput or in high content can be implemented, as derivatives of pluripotent stem cells can be obtained at near homogeneity and are amenable to miniaturization and standardization of cell processes. At I-Stem, we have already exploited this potential for several pathologies that affect the brain, including Huntington's disease, Myotonic Dystrophy type I and Lesch-Nyhan disease. In parallel, functional genomics can also be implemented on large-scale platforms, in a search for yet unknown mechanisms and proteins involved in pathological signaling pathways.

Understanding the mechanisms by which a genetic variation contributes to diseases is a central aim of human genetics and will greatly facilitate the development of preventive strategies and treatments. Recent advances in cell biology have fueled the prospect that the difficulty in unraveling the disease mechanisms may be overcome thanks to the availability of disease-specific pluripotent stem cells. The concept is simple: pluripotent stem cells are capable, by definition, of

M. Peschanski (✉) • C. Martinat
INSERM/UEVE 861, I-Stem, AFM, 5 rue Henri Desbruères, Evry 91000, France
e-mail: mpeschanski@istem.fr

F.H. Gage and Y. Christen (eds.), *Programmed Cells from Basic Neuroscience to Therapy*, Research and Perspectives in Neurosciences 20,
DOI 10.1007/978-3-642-36648-2_9, © Springer-Verlag Berlin Heidelberg 2013

differentiating into most if not all adult cell types. Therefore, they afford access to cell populations that are difficult to obtain otherwise (for example, neurons), and thus they allow investigators to follow the disease progression and to gain valuable insight into its pathophysiology. Another crucial advantage of pluripotent stem cells is their capacity to self-renew, facilitating cell-based genetic or drug screening. Classically, human disease-specific pluripotent stem cells have been generated using two distinct methods. First, disease-specific human embryonic stem cells (hESCs) can be derived from pre-implantation embryos obtained during in vitro fertilization. Second, embryonic-like stem cells can be obtained through the conversion of somatic cells isolated from patients by genetic manipulations (Takahashi et al. 2007; Thomson et al. 1998). Due to their origin, pluripotent stem cells are considered to be inexhaustible, scalable and physiologically native material for experimentation. Pluripotent stem cells represent the genetic background of the original patient, which may be crucial because phenotype variation may entail interactions between a polymorphism and modifier loci.

Disease modeling using human pluripotent stem cells relies on two important conditions: (1) the differentiation of pluripotent stem cells into pathologically relevant populations, and (2) the ability to recapitulate key aspects of the disease in a time frame compatible with in vitro studies. Although neither of these conditions is easily reached, the use of disease-specific pluripotent stem cells to address pathophysiological questions has exploded over the past 5 years. Most of these studies have focused on monogenic diseases but the field also envisions using pluripotent stem cells to investigate complex disorders. Here, we discuss these studies and the insights that they offer.

Modeling Monogenic Diseases with ES Cell Lines Derived from PGD Embryos

Until 5 years ago, there were two ways to obtain disease-specific human pluripotent stem cells: the genetic modification of existing hESCs or the derivation of hESCs from embryos carrying the causal mutation related to a monogenic disease detectable during pre-implantation genetic diagnosis (PGD; see Table 1 for the PGD-embryo cell lines derived by I-Stem in collaboration with IGBMC, Dr. Stéphane Viville). In 2004, Urbach et al. were the first to demonstrate that genetically modified hESCs can reproduce, to some extent, molecular features associated with the Lesch-Nyhan syndrome. In this study, the *hrpt1* gene was mutated in hESCs through homologous recombination. The resulting hESCs showed an absence of *hrpt1* activity and produced more uric acid than unmodified "wild-type" cells (Urbach et al. 2004). However, generating mutant hESCs lines has faced challenges due to the inefficient methods of genetically modifying hESCs.

Table 1 List of pluripotent stem cells derived from embryos carrying a mutant gene responsible for a monogenic disease that have been derived after pre-implantation genetic diagnosis through a collaboration between Stéphane Viville's team (IGBMC Strasbourg, France) and I-Stem. These lines are available upon request, within French regulations

Strasbourg	Pathology	N° Ag/Biomed	Karyotype
STR-I-155-HD	HD (Huntington disease)	FE07-051-L1	46, XX
STR-I-171-GLA	GLA (Fabry disease)	FE08-056-L1	46, X
STR-I-189-FRAXA	FRAXA (Fragile X)	FE08-065-L1	46, XX
STR-233-FRAXA	FRAXA (Fragile X)	FE08-087-L1	46, XY
STR-I-203-CFTR	CFTR (Cystic fibrosis)	FE08-072-L1	46, XX
STR-I-251-CFTR	CFTR (Cystic fibrosis)	FE09-0046-L1	46, XX
STR-I-209-MEN2a	NEM2a (Neoplasia endocrine multiple type 2a)	FE08-075-L1	46, XY
STR-I-211-MEN2a	NEM2a (Neoplasia endocrine multiple type 2a)	FE08-076-L1	47, XY +16
STR-I-221 or 263	SCA2 (Spino-cerebellar ataxia type 2)	FE08-081-L1	46, XX
STR-I-229 or 271 ou 275	MTMX (X linked myotubular myopathy)	FE08-085-L1	46, XY
STR-I-231 or 271 or 275	MTMX (X linked myotubular myopathy)	FE08-086-L1	46, XY
STR-I-301-MFS	MFS (Marfan syndrome)	FE09-071-L1	46, XX
STR-I-305-APC	APC (Adenomatosis polyposis coli)	FE09-073-L1	46, XY
STR-I-315-CMT1a	CMT1a (Charcot-Marie Tooth Type 1a)	FE09-078-L1	46, XY
STR-I-347-FRAXA	FRAXA (Fragile X)	FE09-094-L1	46, X
STR-I-355-APC	APC (Adenomatosis polyposis coli)	FE09-0271-L1	46, XY
STR-I-443-NF1	NF1 (Neurofibromatosis type I)	FE09-0313L1	46, XX
STR-I-441-NF1	NF1 (Neurofibromatosis type I)	FE09-0315L1	46, XY
STR-I-437-NF1	NF1 (Neurofibromatosis type I)	FE09-0316L1	46, XX

By performing PGD on embryos, disease-specific hESCs have been derived for a dozen different monogenic diseases. Up to now, few of these studies have really recapitulated the phenotypes of the diseases.

In the case of Myotonic Dystrophy type I (DM1), our laboratory recently demonstrated that PGD-hESCs can be used to identify new physiopathological mechanisms (Marteyn et al. 2011). DM1 patients suffer from muscle wasting and multiple defects in their central nervous system. Although progress has been made concerning the identification of the mutation implicated in this disease, the molecular mechanisms that underlie the disease, which could be targets for treatment, are not well understood (Lee and Cooper 2009). We first demonstrated that neural cells derived from two different DM1-hESCs lines recapitulated some molecular features associated with DM1, such as the presence of toxic ribonucleoprotein inclusions and the splicing defect of the NMDAR1 as described in patients (Jiang et al. 2004). A genome-wide comparison of DM1 hESCs differentiated into neural cells with their healthy counterparts allowed the identification of a reduced expression of two genes of the SLITRK family in mutant cells that was mirrored in DM1 brain samples. SLITRK proteins are involved in the outgrowth of neurites and the formation of synapses, which are sites of communication between nerve and muscle cells. The mis-expression of *SLITRK* genes in DM1 cells could be related to a defect in DM1 hESCs-derived motoneurons to correctly connect with their target muscular cells, reminiscent of a few observations made in DM1 patients and in a DM1 mouse model. Interestingly, this phenotype can be both rescued by over-expression of missing genes in mutant cells and induced in control cells by knocking down expression of the genes.

Monogenic Diseases Modeling Using Human iPS Cells

Concurrent with the development of disease-specific hESCs lines, the conversion of somatic cells from patients into iPS cells has transformed the prospect for disease modeling (Takahashi et al. 2007; Park et al. 2008). Successful generation of hiPS cells has, for instance, been reported for both monogenic diseases such as epidermolysis bullosa (Tolar et al. 2010) and congenital dyskeratosis (Agarwal et al. 2010) and for more complex conditions such as autism spectrum disorders (Marchetto et al. 2010), schizophrenia (Brennand et al. 2011) and sporadic middle- or late-age onset neurodegenerative diseases such as amyotrophic lateral sclerosis and Parkinson's disease (Park et al. 2008; Dimos et al. 2008).

Up to now, only partial disease modeling has been reported, mostly for monogenic diseases. This modeling can become even more relevant when animal models for the pathology do not recapitulate the exact human situation, as with spinal muscular atrophy (SMA) disorder. Homozygous absence of the survival motoneuron gene *SMN1* is the primary cause of SMA, whereas disease severity is mainly determined by the expression level of the *SMN1* paralog, *SMN2*, which is only present in humans. Ebert et al.(2009) change by Different groups derived hiPS cells from fibroblasts from SMA patients and showed a decrease in hiPS cell-derived motoneuron survival after 6 weeks of differentiation, compared with hiPS cell-derived motoneurons from the unaffected mother

(Ebert et al. 2009; Sareen et al. 2012; Chang et al. 2011). These SMA hIPS cells also respond to compounds known to increase non-specifically *SMN* expression, indicating a potential future drug-screening platform using the iPS cell technology.

Another study by Studer and colleagues (Anderson et al. 2001) on the neurodegenerative familial dysautonomia (FD) disease has also suggested the possibility of using disease-specific hiPS cells for drug screening. Most of the FD patients harbor a point mutation in the IKB kinase complex-associated protein (IKBKAP), leading to a tissue-specific splicing defect with reduced levels of the transcript encoding IKBKAP. hiPS cells derived from three patients with FD revealed that neural crest precursors, specifically, had low levels of *IKBKAP* expression (Lee et al. 2009). In addition, a defect in neuronal differentiation and migration was reported. Testing three previous compounds known to act on the splice and the absolute levels of IKBKAP, kinetin was identified for its ability to reduce the levels of the mutant IKBKAP splice forms. It also improved neuronal differentiation but not cell migration in hiPS cell-derived neuronal differentiation, suggesting incomplete phenotype complementation. However, these findings open the prospect for drug screening using kinetin-like variations with neuronal differentiation and cell migration phenotypes as readout in future studies.

Cell-Based Models for Identification and Testing Therapies

Drug discovery is time consuming and costly. The high failure rate of compounds in clinical development is a major problem for the pharmaceutical industry. Up to 90 % of compounds fail at different steps in clinical trials, due to low efficiency or safety issues. One of the current drug identification methods is based on high-throughput screening (HTS) technologies, allowing the rapid evaluation of millions of compounds. Most of the cellular models used for drug screening are based on genetically modified rodent or immortalized human cell lines containing reporter gene expression systems. However, these cellular models, as well as the use of overexpressed reporter genes, do not necessarily provide an accurate system to fully evaluate the exact effect of the drug on physiological condition. Recent advances in the stem cell field are creating possibilities for high-throughput and high content screening. For example, more than one million compounds have been tested using adult mouse neural stem cells in a 1,536-well plate format to identify possible inducers of proliferation and differentiation (Liu et al. 2009). In 2008, Desbordes et al. described a method to plate hES cells into a 384-well plate for HCS of a small quantity of compounds that were related to self-renewal and differentiation (Desbordes et al. 2008). Recently, this technique has allowed us, in collaboration with Roche, to screen a huge compound library in a search for drugs that would facilitate neurogenesis.

In conclusion, it remains to be determined whether these artificial in vitro models will indeed improve the specificity and the sensitivity that are required to predict responses in a complex in vivo environment.

Conclusions

Among the different applications of pluripotent stem cells, a near-term attainable goal using human pluripotent stem cells will be for pathological modeling and mechanistic studies that should lead to the development of new therapeutic strategies. Disease-specific human pluripotent stem cells are already influencing the way in which disease modeling and development of adapted therapeutic strategy are approached. However, to translate disease-specific iPS cells into clinically informative models for mechanistic studies and therapeutic screening, several technical challenges need to be solved: generation of transgene-free, fully reprogrammed patient iPS cells, optimization and standardization of differentiation methods, development of phenotypic assays relevant to the disease process, and advances in genetic modifications to create isogenic controls. We assume that future studies will be focused on finding solutions to these challenges.

In view of the recent studies demonstrating that pluripotent stem cells can recapitulate, in certain cases, some disease-associated phenotypes, it is clear that, in the near future, an expanding collection will be developed. Identification of expected phenotypes based on previous analyses of animal or cellular models is a pre-requisite to validate the specificity of the models. However, identification of novel mechanisms or cellular phenotypes remains the most exciting and promising opportunity offered by these disease-specific models. The combination of the "omic" technologies and pluripotent stem cells may eventually become the potentially ground-breaking cocktail for this purpose. Monitoring simultaneously a large number of cellular pathways, these technologies should facilitate the identification of signaling molecules involved in functional disturbances, cell damage and damage responses. The association of "omic" technologies with functional dissection, by using the plasticity of pluripotent stem cells, should reveal new pathological mechanisms and help identify new therapeutics.

References

Agarwal S, Loh YH, McLoughlin EM, Huang J, Park IH, Miller JD, Huo H, Okuka M, Dos Reis RM, Loewer S, Ng HH, Keefe DL, Goldman FD, Klingelhutz AJ, Liu L, Daley Q (2010) Telomere elongation in induced pluripotent stem cells from dyskeratosis congenita patients. Nature 464:292–296

Anderson SL, Coli R, Daly IW, Kichula EA, Rork MJ, Volpi SA, Ekstein J, Rubin BY (2001) Familial dysautonomia is caused by mutations of the IKAP gene. Am J Hum Genet 68:753–758

Brennand KJ, Simone A, Jou J, Gelboin-Burkhart C, Tran N, Sangar S, Li Y, Mu Y, Chen G, Yu D, McCarthy S, Sebat J, Gage FH (2011) Modelling schizophrenia using human induced pluripotent stem cells. Nature 473:221–225

Chang T, Zheng W, Tsark W, Bates S, Huang H, Lin RJ, Yee JK (2011) Phenotypic rescue of induced pluripotent stem cell-derived motoneurons of a spinal muscular atrophy patient. Stem Cells 29(12):2090–2093

Desbordes SC, Placantonakis DG, Ciro A, Ciro A, Socci ND, Lee G, Djaballah H, Studer L (2008) High-throughput screening assay for the identification of compounds regulating self-renewal and differentiation in human embryonic stem cells. Cell Stem Cell 2:602–612

Dimos JT, Rodolfa KT, Niakan KK, Weisenthal LM, Mitsumoto H, Chung W, Croft GF, Saphier G, Leibel R, Goland R, Wichterle H, Henderson CE, Eggan K (2008) Induced pluripotent stem cells generated from patients with ALS can be differentiated into motor neurons. Science 321:1218–1221

Ebert AD, Yu J, Rose FF Jr, Mattis VB, Lorson CL, Thomson JA, Svendsen CN (2009) Induced pluripotent stem cells from a spinal muscular atrophy patient. Nature 457:277–280

Jiang H, Mankodi A, Swanson MS, Moxley RT, Thornton CA (2004) Myotonic dystrophy type 1 is associated with nuclear foci of mutant RNA, sequestration of muscleblind proteins and deregulated alternative splicing in neurons. Hum Mol Genet 13:3079–3088

Lee JE, Cooper TA (2009) Pathogenic mechanisms of myotonic dystrophy. Biochem Soc Trans 37 (Pt 6):1281–1286

Lee G, Papapetrou EP, Kim H, Chambers SM, Tomishima MJ, Fasano CA, Ganat YM, Menon J, Shimizu F, Viale A, Tabar V, Sadelain M, Studer L (2009) Modelling pathogenesis and treatment of familial dysautonomia using patient-specific iPSCs. Nature 461:402–406

Liu Y, Lacson R, Cassaday J, Ross DA, Kreamer A, Hudak E, Peltier R, McLaren D, Muñoz-Sanjuan I, Santini F, Strulovici B, Ferrer M (2009) Identification of small-molecule modulators of mouse SVZ progenitor cell proliferation and differentiation through high-throughput screening. J Biomol Screen 14:319–329

Marchetto MC, Carromeu C, Acab A, Yu D, Yeo GW, Mu Y, Chen G, Gage FH, Muotri AR (2010) A model for neural development and treatment of Rett syndrome using human induced pluripotent stem cells. Cell 143:527–539

Marteyn A, Maury Y, Gauthier MM, Lecuyer C, Vernet R, Denis JA, Pietu G, Peschanski M, Martinat C (2011) Mutant human embryonic stem cells reveal neurite and synapse formation defects in type 1 myotonic dystrophy. Cell Stem Cell 8:434–444

Park IH, Arora N, Huo H, Maherali N, Ahfeldt T, Shimamura A, Lensch MW, Cowan C, Hochedlinger K, Daley GQ (2008) Disease-specific induced pluripotent stem cells. Cell 134:877–886

Sareen D, Ebert AD, Heins BM, McGivern JV, Ornelas L, Svendsen CN (2012) Inhibition of apoptosis blocks human motor neuron cell death in a stem cell model of spinal muscular atrophy. PLoS One 7(6):e39113

Takahashi K, Tanabe K, Ohnuki M, Narita M, Ichisaka T, Tomoda K, Yamanaka S (2007) Induction of pluripotent stem cells from adult human fibroblasts by defined factors. Cell 131:861–872

Thomson JA, Itskovitz-Eldor J, Shapiro SS, Waknitz MA, Swiergiel JJ, Marshall VS, Jones JM (1998) Embryonic stem cell lines derived from human blastocysts. Science 282:1145–1147

Tolar J, Xia L, Riddle Mj MJ, Lees CJ, Eide CR, McElmurry RT, Titeux M, OsbornMJ LTC, Hovnanian A, Wagner JE, Blazar BR (2010) Induced pluripotent stem cells from individuals with recessive dystrophic epidermolysis bullosa. J Invest Dermatol 131:848–856

Urbach A, Schuldiner M, Benvenisty N (2004) Modeling for Lesch–Nyhan disease by gene targeting in human embryonic stem cells. Stem Cells 22:635–641

Modeling Autism Spectrum Disorders Using Human Neurons

Alysson Renato Muotri

Abstract The cellular and molecular mechanisms of neurodevelopmental conditions such as autism spectrum disorders (ASDs) have been studied intensively for decades. The unavailability of live patient neurons for research, however, has represented a major obstacle in the elucidation of the disease etiologies. The generation of human neurons from induced pluripotent stem cells (iPSCs) derived from patients with ASDs offers a novel and complementary opportunity for basic research and the development of therapeutic compounds aiming to revert or ameliorate the condition. The findings of relevant phenotypes in Rett syndrome iPSC-derived neurons suggest that iPSC technology offers a novel and unique opportunity for understanding and developing therapeutics for other ASDs. Neurons-in-a-dish from syndromic forms of ASD open new avenues for the stratification of different subtypes of idiopathic autism. In this chapter, I will discuss the conceptual and practical issues related to modeling ASD using human neurons.

Introduction

Science has improved human life and the understanding of human disease by taking advantage of models to mimic several conditions in the laboratory. Models are simplified representations or reflections of the reality. Thus, all models are useful in certain situations. Models are inaccurate by nature and all models have specific intrinsic limitations, but the best models allow the complexity of a system to approach the complexity of a human disease. Autism Spectrum Disorders (ASDs)

A.R. Muotri (✉)
Department of Pediatrics/Rady Children's Hospital San Diego, Department of Cellular & Molecular Medicine, Stem Cell Program, University of California San Diego, School of Medicine, 2880 Torrey Pines Scenic Road – Sanford Consortium, Room 3005, La Jolla CA 92093, MC 0695, USA
e-mail: muotri@ucsd.edu

F.H. Gage and Y. Christen (eds.), *Programmed Cells from Basic Neuroscience to Therapy*, Research and Perspectives in Neurosciences 20,
DOI 10.1007/978-3-642-36648-2_10, © Springer-Verlag Berlin Heidelberg 2013

are complex neuropsychiatric conditions, involving multiple genetic targets across several neural circuits in the brain. The lack of usable neuronal samples from post-mortem brains and the inability to isolate populations of neurons from living subjects have blocked progress toward studying the underlying cellular and molecular mechanisms of ASDs. Studies of cadaver tissue are problematic in developmental disorders because disease onset usually precedes death by decades. Frozen tissue sections are of limited use for studying cellular physiology and neural networks. Most of the time, the tissue is not well-preserved and even information about gene expression or anatomy can be lost due to inappropriate handling. Additionally, the long-term medication history of patients could affect the observed phenotype in the tissue. Peripheral tissues, such as blood or skin, have been extremely helpful to the understanding of ASD genetics but are not suitable to follow up with relevant biological questions. A similar case can be seen in computational methods, which are thus far restricted to data collected from peripheral tissue. Brain imaging allows you to study circuitries but only at a very low magnification, under the influence of the environment, and it also has limited experimental power. Finally, animal models often do not recapitulate more than a few aspects of complex human diseases, if at all, which has been particularly problematic in the case of ASDs. The lack of ASD-like behaviors in several knockout mouse models, based upon knowledge of genes related to ASDs, reflects the inherent differences between the two species' genetic backgrounds and neural circuits. In fact, while there are multiple genetic mutations that disrupt social behavior in mice, the vast majority do not appear to have direct relevance to ASDs. Conversely, many ASD mutations have no effect in mice or lead to phenotypes that do not mimic the human disease (Silverman et al. 2010). These observations illustrate the challenges associated with complex neuropsychiatric modeling in animals and future translation into human therapies (Dragunow 2008). Thus, the ASD field lacks an appropriate human model and would greatly benefit from unlimited supplies of neurons so that experiments could be performed in controlled situations.

Unlimited Neuronal Potential from Human Pluripotent Stem Cells

Pluripotent human embryonic stem cells (hESCs) have been successfully isolated from early stage human embryos (blastocysts), can self-renew, and differentiate into various cell types, offering an unlimited source of cell types for research (Thomson et al. 1998). However, due to ethical and moral reasons, it is not possible to demonstrate that hESCs can actually contribute to different cell types and tissues in a real person. Perhaps the most rigorous demonstration that hESCs could actually become functional human neurons, fully integrated into the neuronal network, was performed after successful transplantation of hESCs in the ventricles of embryonic mouse brains (Muotri et al. 2005b). These "chimeric brains" carried human cells that were differentiated into functional neural lineages and generated mature, active human neurons that successfully integrated into the adult mouse forebrain. A small fraction of the transplanted cells integrated individually or in small clusters into the host tissue with similar morphometric dimensions as adjacent host cells, including shape, size and orientation, and adjusted to the pre-existing cellular architecture. Transplanted cells co-localized with markers specific for neuronal subtypes. Evidence of synaptic inputs was apparent in the presence of arborized dendrites with spines, suggesting that glutamate-containing terminals contacted these dendrites. Ultrastructural analysis also confirmed that human cells received synaptic input and exhibited mature features, such as pools of presynaptic vesicles adjacent to a postsynaptic density. Moreover, transplanted cells showed neuronal properties similar to neurons under comparable electrophysiological recording conditions. Such observations, plus other evidence in vitro, made a convincing argument that human pluripotent stem cells could actually form functional neurons and thus function as a model for early stages of brain development.

However, to develop cellular models of human disease, it is necessary to generate new cell lines with genomes that are pre-disposed to diseases. By taking advantage of pre-implantation genetic diagnosis, it was possible to generate hESCs carrying mutations in specific genes known to cause human diseases. This procedure was conducted with cystic fibrosis, Huntington's disease, Marphan syndrome, Fragile-X, and other monogenetic diseases (Bradley et al. 2011; Mateizel et al. 2006; Pickering et al. 2005; Verlinsky et al. 2005). Forward genetics was also used to generate hESC disease models by homologous recombination. Perhaps the first example was generated by the Benvenisty group, which used gene targeting to knockout the *HPRT1* gene, responsible for Lesch-Nyham syndrome (Urbach et al. 2004). Unfortunately, apart from the ethical and political concerns related to hESC line derivation, this strategy is also limited by the availability of human blastocysts and by the number of genes one can manipulate in hESCs –notoriously resilient for gene targeting (Giudice and Trounson 2008). Complex disorders, where multiple genes are affected, or "sporadic" diseases such as ASD, schizophrenia or amyotrophic lateral sclerosis (ALS) in which the genetic alteration is not previously known, cannot be modeled using forward genetics in hESCs.

The reprogramming technology provides a possible solution to this problem as it allows the genomes of human individuals afflicted with ASD to be captured in a pluripotent stem cell line. Recently, reprogramming of somatic cells to a pluripotent state by over-expression of specific genes was accomplished using mouse fibroblasts (Takahashi and Yamanaka 2006). To reprogram somatic cells, Takahashi and Yamanaka tried 24 genes that were previously demonstrated to be expressed in ESCs for their ability to induce somatic cells into ESC-like cells. Surprisingly, they found that only four retroviral-mediated transcription factors, the octamer binding protein 4 (OCT4, also known as POU5F1), SOX2, Krüppel-like factor 4 (KLF4) and MYC, were sufficient to jump start the expression of endogenous pluripotency genes in somatic cells. Despite the fact that the biology of reprogramming is not completely yet understood, it is clear that the repression of gene expression by the binding of transcription factors and epigenetic marks in the chromatin observed in donor somatic cells can be reversed by reprogramming factors to developmentally regress the cells to an earlier state (Ho et al. 2011). These reprogrammed cells were able to form embryoid bodies in vitro and teratomas in vivo and contributed to several tissues in chimeric embryos when injected into mouse blastocysts. The report of human reprogrammed cells using the same set of transcriptional factors appeared soon after (Takahashi et al. 2007; Yu et al. 2007). These cells, named induced pluripotent stem cells (iPSCs) can be derived from cells isolated from peripheral tissues of normal individuals or of people affected by several conditions.

iPSCs and hESCs are very similar, but significant differences can be detected when comparing them at higher magnification (Marchetto et al. 2009). Gene expression differences between iPSCs and hESCs can be caused by incomplete silencing of genes expressed in donor cells and failure to fully induce pluripotent genes in reprogrammed cells, likely reflecting incomplete resetting of somatic expression (Chin et al. 2009). Epigenetic markers also seem to differ between the two types of pluripotent cells. Hotspots of aberrant epigenomic reprogramming were reported in methylated regions proximal to centromeres and telomeres in iPSCs (Lister et al. 2011). Besides epigenetic modifications, genetic alterations can also occur during the reprogramming process. Sometimes iPSC lines display abnormal karyotypes (Mayshar et al. 2010) and large copy number variations (CNVs; Laurent et al. 2011). Interestingly, some of these CNV alterations tend to disappear after several passages of the cells, probably due to selection in culture. Nonetheless, extensive genetic and epigenetic assessments should become a standard procedure to identify the truly reprogrammed cells from those that are only partially reprogrammed or unstable and to ensure the quality of iPSCs used for experiments.

Isogenic pluripotent cells are attractive not only for their potential therapeutic use with lower risk of immune rejection but also for their prospects to further understanding of complex diseases with heritable and sporadic conditions (Marchetto et al. 2010b; Muotri 2009). Such cells can then be differentiated to human neurons to evaluate whether the captured genome alters cellular phenotype in a similar manner as predicted by mechanistic models of ASDs (Fig. 1). An iPSC model may also address human-specific effects and avoid some aspects of the well-known limitations of animal models, such as the absence of a human genetic background. Although iPSCs have been generated for several neurological diseases (Marchetto et al. 2011) the demonstration of disease-specific pathogenesis and phenotypic rescue in relevant cell types is a current challenge in the field, with only a handful of proof-of-principle examples to date, including spinal muscular atrophy (Ebert et al. 2009), Down syndrome (Baek et al. 2009), Rett syndrome (Marchetto et al. 2010a), schizophrenia (Brennand et al. 2011) and others (Grskovic et al. 2011; Robinton and Daley 2012; Saporta et al. 2011).

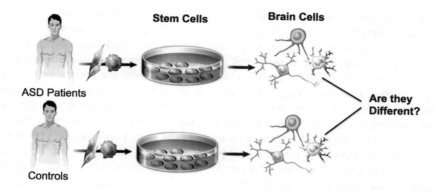

Fig. 1 Modeling ASD with induced pluripotent stem cells (iPSCs). Peripheral cells, isolated from skin or dental pulp, can be reprogrammed to a pluripotent state and propagated in large amounts. These cells can be coaxed to differentiate into brain cells, including distinct neuronal subtypes for further phenotypic evaluation. If differences are observed between a cohort of ASD and control neurons, these alterations can be used as readouts in drug-screening platforms

Potential Limitations for Disease Modeling

As with other models of ASDs, the iPSC system has important limitations. The mechanisms behind cellular reprogramming are currently unknown. Understanding the current pitfalls of this technology is crucial to making correct interpretations and plausible extrapolations to the human brain. As mentioned before, some regions of the genome may not be completely reprogrammed. The implications of the existence of epigenetically resilient regions of the genome to disease modeling were demonstrated by comparing hESCs and iPSCs as models for Fragile X (Urbach et al. 2010). Fragile X is a common form of mental retardation characterized by a lack of expression of *FMR1*, a gene that is normally expressed in hESCs but is prompted to silence during differentiation. Interestingly, the mutant FMR locus in iPSCs derived from Fragile-X patients is not epigenetically reset during the reprogramming process, an important difference between the hESC and iPSC models.

Cells in culture represent an artifact; they are not in the exact same environment as they would be in vivo. They are missing important signaling pathways, interaction with other cells, and the holistic environment of different tissues in a living organism. Moreover, our culture conditions for maintenance, propagation and differentiation of iPSCs are not optimized but were achieved based on previous data from mice. Thus, it is possible that important signaling information is missing or over-stimulated in the culture system, masking potential cellular phenotypes. Additional limits to the neural conversion of iPSCs are the lower efficiency and higher variability of neural differentiation in iPSCs compared to ESC lines (Hu et al. 2010) and the existence of intra-individual variation within different clones from the same individual.

Another challenge in the disease-in-a-dish field is the derivation of relevant neuronal subtypes. In theory, pluripotent stem cells can be differentiated in all neuronal types of the human brain (Muotri and Gage 2006). Practically, there are only a few protocols to induce iPSC differentiation into specific subtypes of neurons. The differentiation usually contains a heterogeneous population of cell types, such as astrocytes, oligodendrocytes, or even non-neuronal cell types, and the relevant neuronal subtypes need to be sorted out or visualized using specific reporter genes. The use of the synapsin promoter driving EGFP has been instrumental in visualizing and sorting neurons from the mixed population (Marchetto et al. 2010a). However, the synapsin promoter is not specific to any neuronal subtypes and characterization of promoter-specific markers will be necessary to isolate these cells in the future. The different types of neurons can then be characterized by their morphology, gene expression and electrical activity to demonstrate their specificity, maturity and connectivity.

While it is clear what type of neuron is mostly affected in some neurological diseases, this is not the case with ASDs. Here, I would argue that the iPSC system could be used as a toolbox to help determine the impact of ASD in different neuronal types. It is the only model that allows progressive time-course analyses of the different neuronal types. It will be possible to investigate the precise neuronal types that are affected in ASD and to elucidate the cellular and molecular defects that contribute to disease initiation and progression. As the protocols become more robust, one will be able to systematically differentiate neurons from distinct brain regions to look for phenotypes. This strategy will provide us with insights into timing and neuron-specific information about early stages of the disease process.

But having the relevant neuronal type in culture does not guarantee that disease neurons will behave differently than controls. It is possible that non-cell autonomous effects, such as different cell types, three-dimensional scaffolding, or maturation timing, may also contribute to neuronal phenotypes. In that case, it is expected that neurodevelopmental phenotypes will be easier to spot than phenotypes in late-onset diseases. The former may require some external stimuli, such as the presence of stressors, to reveal the differences in patient-derived neurons.

Another important limitation is the use of appropriate controls. Intuitively, for well-characterized monogenetic diseases, the ideal controls would be ones that differ from the patient only in the genetic defect. Efforts in this direction can be achieved by manipulation of the iPSCs to introduce genetic mutations in control cell lines or to restore the mutation from a patient cell line (Liu et al. 2011; Soldner et al. 2011). Another strategy to generate "isogenic" cell lines is to take advantage of X-inactivation in female cell lines. Due to the fast X inactivation process during reprogramming, it is possible to generate iPSC cell clones carrying the mutant or the wild-type version of an X-linked affected gene. This strategy was used to model Rett syndrome, which affects female patients with mutations in the X-linked MeCP2 gene (Cheung et al. 2011; Marchetto et al. 2010a). However, it is important to consider that even "isogenic" clones in culture will accumulate mutations in their genome over time and, thus, there will never be an ideal control line. Although the implications of these alterations may be small, we should not underestimate the selection process going on in a dish. For non-monogenetic diseases, or when the mutations are not known, such as sporadic autism, the challenge is even bigger. Variations between cell lines and even between iPSC clones from the same individual can influence the phenotypic readout. In that case, a large cohort of well-characterized control cell lines is invaluable. Real phenotypes could be identified when the variations between controls and patients are significantly higher than the intrinsic variations inside each group. Unfortunately, the generation and characterization of individual iPSC clones is expensive and time-consuming, restricting the number of cell lines that an individual can handle. A possible useful strategy for these types of diseases is the coordination of consortium initiatives, where multiple laboratories would contribute to the pool of different cell types and development of phenotypic assays. Consortium initiatives could also be useful in creating banks of genetically characterized controls that could be used to research the closest controls to pathological cases.

A final challenge for the iPSC model is the validation of phenotypes observed in human neurons to show that this model can recapitulate the disease in a dish. Comparison with postmortem brain tissues is perhaps the most obvious step towards validation; however, the lack of consistency does not mean that the phenotype is not valid. Important neuronal alterations during development may not be present in adult tissues due to brain compensation, for instance. Validation in animals is an attractive alternative and may reveal important conserved neural pathways/circuits between the two species. But again, a negative correlation with mouse models does not imply the phenotypes are not important for humans. Moreover, in the case of sporadic ASDs, where animal models offer limited information about the human brain and there is not a large amount of data describing phenotypic variations in neuroanatomical circuits and molecular pathways, validation can be problematic.

Early Insights into ASD Neurons Derived from iPSCs

The use of monogenetic forms of ASD was wisely chosen as proof-of-principle that neurons derived from these patients could recapitulate important aspects of the *disease* in vitro. Studies of single gene mutations accelerated the discovery of causal mechanisms related to neuronal phenotypes. Monogenic disorder modeling provides the opportunity to perform gain- and loss-of-function experiments to confirm that the phenotypes observed are disease, specific as opposed to a general, non-specific effect. These models can bring new insights to other forms of ASDs. Moreover, by capturing the genetic heterogeneity of ASDs in a pluripotent state, the iPSC model has the potential to determine whether patients carrying distinct mutations in disparate genes share common cellular and molecular neuronal phenotypes.

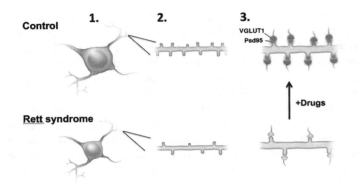

Fig. 2 Examples of neuronal differences observed in Rett syndrome (RTT) neurons compared to controls. Neurons derived from RTT-iPSCs showed reduced soma size, low density of dendritic spines and defects in glutamatergic synapses. Some of these defects were shown to be rescued with candidate drugs, as a proof-of-principle that human RTT neurons can be reverted and behave like neurons derived from non-affected control individuals. The red (VGLUT1) and the green (Psd95) puncta represent markers for glutamatergic connections

To date, two syndromic ASDs have been modeled using the iPSC strategy. We recently demonstrated the utility of iPSCs to investigate the functional consequences of mutations in the gene encoding the Methyl CpG binding protein-2 (MeCP2) on neurons from patients with Rett syndrome (RTT), a syndromic form of ASD (Marchetto et al. 2010a). Neurons derived from RTT-iPSCs carrying four different MeCP2 mutations showed several alterations compared to five healthy non-affected individuals, such as decreased soma size, altered dendritic spine density and reduced excitatory synapses (Fig. 2). Importantly, these phenotypes were validated using wild-type MeCP2 cDNA and specific shRNAs against MeCP2 in gain- and-loss-of-function experiments. Some of these cellular defects were immediately validated by independent groups, revealing the robustness and reproducibility of the system (Ananiev et al. 2011; Cheung et al. 2011; Kim et al. 2011). We were able to rescue the defects in a number of glutamatergic synapses using two candidate drugs, insulin growth factor 1 (IGF1) and gentamicin. IGF1 is considered to be a candidate for pharmacological treatment of RTT and, potentially, other CNS disorders in ongoing clinical trials (Tropea et al. 2009). Gentamicin, a read-through drug, was also used to rescue neurons carrying a nonsense MeCP2 mutation, by elevating the amount of MeCP2 protein. These observations provide valued information about RTT and, potentially, other ASD patients, since they suggest that pre-symptomatic defects may represent novel biomarkers to be exploited as diagnostic tools. The data also suggest that early intervention may be beneficial.

Moreover, we took advantage of the RTT-iPSCs to demonstrate that neural progenitor cells carrying MeCP2 mutations have increased susceptibility for L1 retrotransposition. Long interspersed nuclear elements-1 (LINE-1 or L1s) are abundant retrotransposons that comprise approximately 20 % of mammalian genomes (Gibbs et al. 2004; Lander et al. 2001; Waterston et al. 2002) and are highly active in the nervous system (Coufal et al. 2009; Muotri et al. 2005a). Our data demonstrate that L1 retrotransposition can be controlled in a tissue-specific manner and that disease-related genetic mutations can influence the frequency of neuronal L1 retrotransposition (Muotri et al. 2010). This work revealed an unexpected and novel phenomenon, adding a new layer of complexity to the understanding of genomic plasticity, and it may have direct implications for ASDs.

Timothy syndrome (TS) is caused by a point mutation in the CACNA1C gene, encoding the alpha1 subunit of $Ca_v1.2$ protein (Splawski et al. 2004). There are only a few dozen people in the world with TS. Patients suffering from TS often display cardiac arrhythmia, hypoglycemia and global developmental delay. Some individuals are also diagnosed with ASD (Splawski et al. 2004, 2006). To investigate this syndromic, rare form of ASD, the Dolmetsch group derived iPSCs from two TS patients and four controls and then coaxed these cells to differentiate into precursor cells and neurons. The protocol used favored differentiation into cortical neurons. TS-derived neurons showed abnormal calcium signaling, leading to differences in gene expression. The data also suggest that TS-iPSCs produce fewer callosal projection neurons. Moreover, TS neurons produced more catecholamines than neurons derived from healthy individuals (Pasca et al. 2011). Catecholamines, such as norepinephrene and dopamine, are important for sensory gating and social behavior, suggesting an important role in the pathophysiology of ASDs. The excess of catecholamines could be rescued by treating TS neurons with roscovitine, a cyclin-dependent kinase inhibitor that blocks calcium influx across cell membranes. This experiment suggests a potential novel clinical intervention. Interestingly, while many of the differences between control and TS neurons could be recapitulated in a TS mouse model, the excess of catecholamines could not. One interpretation is that species-specific differences in gene regulation may affect the cellular phenotypes associated with TS patients. Alternatively, the differences that could not be recapitulated in mice may have a less important role in the disease. In the end, differences that provide a therapeutic lead could be transferred to clinical trials for future validation.

Moving Towards Modeling Idiopathic ASDs Using iPSCs

Based on the examples of RTT and TS, it is possible to conclude that functional studies using iPSC-derived neuronal cultures of ASD patients can be an important addition to the exploration of the contribution of rare variants to ASD etiology. The notion that very rare mutations may point to key etiological pathways and mechanisms has been repeatedly demonstrated for a wide range of common human disorders, such as Alzheimer's Disease (Scheuner et al. 1996) and hypertension (Ji et al. 2008).

Similarly, a rapidly increasing number of ASD risk regions have recently been identified and there is now considerable effort focused on moving from gene discovery to an understanding of the biological substrates influenced by these various mutations (Bozdagi et al. 2010; Gilman et al. 2011; Gutierrez et al. 2009; Levy et al. 2011; Sanders et al. 2011). While the allelic architecture of ASDs is still being clarified, there is already definitive evidence for a high degree of locus heterogeneity and a contribution by rare and de novo variants (El-Fishawy and State 2010; Geschwind 2011). However, the demonstration of a causal role for these low frequency variants is challenging, particularly as many such mutations defy Mendelian expectations and carry only intermediate risks (Bucan et al. 2009; State and Levitt 2011; Weiss et al. 2008). The sample sizes necessary to demonstrate these types of effects may be impractically large. Moreover, the use of rodent models to evaluate complex human behaviors presents considerable obstacles (Silverman et al. 2010), and patient brain tissue is typically unavailable, requiring the use of peripheral tissues (i.e., blood) for biological studies, with all their inherent limitations as previously discussed. The lack of relevant human-derived cellular models to study ASDs has represented an important obstacle in the effort to link genetic alterations to molecular mechanisms and complex behavioral and cognitive phenotypes. Thus, the use of iPSCs offers an important alternative strategy to investigate the functional consequences of genetic alterations in human neurons in vitro. Creation of large numbers of iPSC lines from patients with idiopathic ASDs could be a novel approach to identify common mechanisms and pathways within cases with different genetic backgrounds.

Merging new sequencing techniques and cellular reprogramming may allow us to characterize alterations in the genome of thousands of ASD patients and to link specific genetic alterations to cellular and molecular phenotypes. At the moment, this approach may be cost prohibitive but both iPSCs and sequencing costs will probably drop exponentially in the near future. Nonetheless, methods for generating neurons on a large scale and automated phenotypic analyses will become essential to take this idea off the ground. The stratification of subtypes of autisms will allow us to recognize molecular pathways that are altered in each individual.

Human Neurons in Drug-Screening Platforms

Unlike in cardiovascular diseases or cancer, in mental disorders it is not possible to compare a set of screened molecules to drugs that have previously demonstrated some benefit in humans. In cancer, for example, it is possible to isolate a biopsy of the tumor and test several drugs for efficiency and toxicity before applying it to the patient. In contrast, live human neurons are not available for drug screening thus far. The lack of drugs in brain disease is a reflection of this missing human model. The studies performed in RTT and TS highlighted the potential of iPSC models in toxicology and drug-screening. Even better, the IGF1 overcorrection observed in some RTT neurons (Marchetto et al. 2010a) indicates that the iPSC technology cannot only recapitulate some aspects of a genetic disease but can also be used to better design and anticipate results from translational medicine. This cellular model has the potential to lead to the discovery of new compounds to treat different forms of ASD.

Drug-screening platforms require "screenable" robust phenotypes in target cell types, such as iPSC-derived neurons. The neuronal differentiation strategies reported to date are not capable of providing vast numbers of homogeneous subtypes of neurons in a reliable, reproducible, and cost-effective fashion. While it seems possible to develop scale up methods for neuronal differentiation in the near future, cellular read-outs are already coming from pioneer published work. Cellular morphology, such as soma size or dendritic spine density, can be measured using high-content imaging software. This might be the case for RTT neurons, where morphological aspects of iPSC-derived neurons could be reproduced by independent laboratories. Early biochemical and gene expression read-outs could be an alternative. However, late read-outs, such as electrophysiological records, may not be ideal due to the time in culture needed to reach proper neuronal maturation. It is certainly possible to use stressors or other environmental agents to speed up the neuronal maturation or enhance the differences between control and patient groups. However, the nature of these agents is still unknown for ASD. An alternative solution may emerge from the direct conversion of neurons from peripheral cells, skipping the pluripotent state (Pang et al. 2011; Vierbuchen et al. 2011). This technology may be faster than neurons generated from iPSCs but it is currently inefficient in humans, is difficult to scale up and has the disadvantage that it does not mimic neuronal development (Vierbuchen and Wernig 2011). Although more robust protocols for human neuronal direct conversion are expected to be reported in the near future, the strategy may not be suitable for neurodevelopmental diseases. It is likely that direct conversion would bypass the developmental period in which ASD phenotype can be observed. The use of direct conversion to produce suitable amounts of neurons for high-throughput screening is limited by the fact that cells produced using direct conversion are not self-renewed. Large amounts of starting material (fibroblasts or other peripheral somatic cells) are therefore required to produce large amounts of neurons. Fibroblasts, for instance, have a finite capacity for replication and cannot be expended indefinitely.

Finally, read-outs need to be suitable for the high-throughput instrumentation screening for drug discovery. More scalable assays will allow characterization of increased numbers of control and patient neurons. There are several chemical libraries that could be used in ASD neurons. It makes sense to use small molecules that cross the blood–brain barrier and have good penetration in the brain. Drug repositioning is a fantastic opportunity. Although this strategy may face some intellectual property challenges, repurposed drugs can bypass much of the early cost and time needed to bring a drug to market. I am very optimistic that the in vitro system using human neurons will accelerate the discovery of novel drugs for RTT and other ASD.

Future Perspectives on the Use of iPSC to Model ASDs

The iPSC strategy is a novel and complementary approach to model ASDs. This new model has the capacity to unify data generated from brain imaging, animal work, and genetics, generating downstream hypotheses that can be tested in well-controlled experiments with relevant cell types. Future work should take advantage of better-characterized ASD cohorts, with well-defined clinical endophenotypes, pharmacological history, and genetic predisposition. By generating hiPSC-derived neurons from these ASD cohorts, one can test whether the clinical outcome is predictive of the magnitude of cellular phenotype, if specific mutations correlate to gene expression differences or if clinical pharmacological response is predictable by human neuronal drug response. In the future, this strategy will give us diagnostic tools to group individual patients into a specific class of autism and to make predictions about whether certain drugs will be beneficial or not. Reproducible and robust ASD neuronal phenotypes can be achieved by intense collaborative consortiums involving several independent laboratories sharing the same patient cohorts, for instance.

Acknowledgments The work in the Muotri lab is supported by grants from the California Institute for Regenerative Medicine (CIRM) TR2-01814, the Emerald Foundation and the National Institutes of Health through the NIH Director's New Innovator Award Program, 1-DP2-OD006495-01.

References

Ananiev G, Williams EC, Li H, Chang Q (2011) Isogenic pairs of wild type and mutant induced pluripotent stem cell (iPSC) lines from Rett syndrome patients as in vitro disease model. PLoS One 6:e25255

Baek KH, Zaslavsky A, Lynch RC, Britt C, Okada Y, Siarey RJ, Lensch MW, Park IH, Yoon SS, Minami T, Korenberg JR, Folkman J, Daley GQ, Aird WC, Galdzicki Z, Ryeom S (2009) Down's syndrome suppression of tumour growth and the role of the calcineurin inhibitor DSCR1. Nature 459:1126–1130

Bozdagi O, Sakurai T, Papapetrou D, Wang X, Dickstein DL, Takahashi N, Kajiwara Y, Yang M, Katz AM, Scattoni ML, Harris MJ, Saxena R, Silverman JL, Crawley JN, Zhou Q, Hof PR, Buxbaum JD (2010) Haploinsufficiency of the autism-associated Shank3 gene leads to deficits in synaptic function, social interaction, and social communication. Mol Autism 1:15

Bradley CK, Scott HA, Chami O, Peura TT, Dumevska B, Schmidt U, Stojanov T (2011) Derivation of Huntington's disease-affected human embryonic stem cell lines. Stem Cells Devel 20:495–502

Brennand KJ, Simone A, Jou J, Gelboin-Burkhart C, Tran N, Sangar S, Li Y, Mu Y, Chen G, Yu D, McCarthy S, Sebat J, Gage FH (2011) Modelling schizophrenia using human induced pluripotent stem cells. Nature 473:221–225

Bucan M, Abrahams BS, Wang K, Glessner JT, Herman EI, Sonnenblick LI, Alvarez Retuerto AI, Imielinski M, Hadley D, Bradfield JP, Kim C, Gidaya NB, Lindquist I, Hutman T, Sigman M, Kustanovich V, Lajonchere CM, Singleton A, Kim J, Wassink TH, McMahon WM, Owley T, Sweeney JA, Coon H, Nurnberger JI, Li M, Cantor RM, Minshew NJ, Sutcliffe JS, Cook EH,

Dawson G, Buxbaum JD, Grant SF, Schellenberg GD, Geschwind DH, Hakonarson H (2009) Genome-wide analyses of exonic copy number variants in a family-based study point to novel autism susceptibility genes. PLoS Gen 5:e1000536

Cheung AY, Horvath LM, Grafodatskaya D, Pasceri P, Weksberg R, Hotta A, Carrel L, Ellis J (2011) Isolation of MECP2-null Rett Syndrome patient hiPS cells and isogenic controls through X-chromosome inactivation. Hum Mol Genet 20:2103–2115

Chin MH, Mason MJ, Xie W, Volinia S, Singer M, Peterson C, Ambartsumyan G, Aimiuwu O, Richter L, Zhang J, Khvorostov I, Ott V, Grunstein M, Lavon N, Benvenisty N, Croce CM, Clark AT, Baxter T, Pyle AD, Teitell MA, Pelegrini M, Plath K, Lowry WE (2009) Induced pluripotent stem cells and embryonic stem cells are distinguished by gene expression signatures. Cell Stem Cell 5:111–123

Coufal NG, Garcia-Perez JL, Peng GE, Yeo GW, Mu Y, Lovci MT, Morell M, O'Shea KS, Moran JV, Gage FH (2009) L1 retrotransposition in human neural progenitor cells. Nature 460:1127–1131

Dragunow M (2008) The adult human brain in preclinical drug development. Nat Rev Drug Discov 7:659–666

Ebert AD, Yu J, Rose FF Jr, Mattis VB, Lorson CL, Thomson JA, Svendsen CN (2009) Induced pluripotent stem cells from a spinal muscular atrophy patient. Nature 457:277–280

El-Fishawy P, State MW (2010) The genetics of autism: key issues, recent findings, and clinical implications. Psychiatr Clin North Am 33:83–105

Geschwind DH (2011) Genetics of autism spectrum disorders. Trends Cogn Sci 15:409–416

Gibbs RA, Weinstock GM, Metzker ML, Muzny DM, Sodergren EJ, Scherer S, Scott G, Steffen D, Worley KC, Burch PE, Okwuonu G, Hines S, Lewis L, DeRamo C, DelgadoO, Dugan-Rocha S, Miner G, Morgan M, Hawes A, Gill R, Celera, Holt RA, Adams MD, Amanatides PG, Baden-Tillson H, Barnstead M, Chin S, Evans CA, Ferriera S, Fosler C, Glodek A, Gu Z, Jennings D, Kraft CL, Nguyen T, Pfannkoch CM, Sitter C, Sutton GG, Venter JC, Woodage T, Smith D, Lee HM, Gustafson E, Cahill P, Kana A, Doucette-Stamm L, Weinstock K, Fechtel K, Weiss RB, Dunn DM, Green ED, Blakesley RW, Bouffard GG, De Jong PJ, Osoegawa K, Zhu B, Marra M, Schein J, Bosdet I, Fjell C, Jones S, Krzywinski M, Mathewson C, Siddiqui A, Wye N, McPherson J, Zhao S, Fraser CM, Shetty J, Shatsman S, Geer K, Chen Y, Abramzon S, Nierman WC, Havlak PH, Chen R, Durbin KJ, Egan A, Ren Y, Song XZ, Li B, Liu Y, Qin X, Cawley S, Worley KC, Cooney AJ, D'Souza LM, Martin K, Wu JQ, Gonzalez-Garay ML, Jackson AR, Kalafus KJ, McLeod MP, Milosavljevic A, Virk D, Volkov A, Wheeler DA, Zhang Z, Bailey JA, Eichler EE, Tuzun E, Birney E, Mongin E, Ureta-Vidal A, Woodwark C, Zdobnov E, Bork P, Suyama M, Torrents D, Alexandersson M, Trask BJ, Young JM, Huang H, Wang H, Xing H, Daniels S, Gietzen D, Schmidt J, Stevens K, Vitt U, Wingrove J, Camara F, Mar Albà M, Abril JF, Guigo R, Smit A, Dubchak I, Rubin EM, Couronne O, Poliakov A, Hübner N, Ganten D, Goesele C, Hummel O, Kreitler T, Lee YA, Monti J, Schulz H, Zimdahl H, Himmelbauer H, Lehrach H, Jacob HJ, Bromberg S, Gullings-Handley J, Jensen-Seaman MI, Kwitek AE, Lazar J, Pasko D, Tonellato PJ, Twigger S, Ponting CP, Duarte JM, Rice S, Goodstadt L, Beatson SA, Emes RD, Winter EE, Webber C, Brandt P, Nyakatura G, Adetobi M, Chiaromonte F, Elnitski L, Eswara P, Hardison RC, Hou M, Kolbe D, Makova K, Miller W, Nekrutenko A, Riemer C, Schwartz S, Taylor J, Yang S, Zhang Y, Lindpaintner K, Andrews TD, Caccamo M, Clamp M, Clarke L, Curwen V, Durbin R, Eyras E, Searle SM, Cooper GM, Batzoglou S, Brudno M, Sidow A, Stone EA, Venter JC, Payseur BA, Bourque G, López-Otín C, Puente XS, Chakrabarti K, Chatterji S, Dewey C, Pachter L, Bray N, Yap VB, Caspi A, Tesler G, Pevzner PA, Haussler D, Roskin KM, Baertsch R, Clawson H, Furey TS, Hinrichs AS, Karolchik D, Kent WJ, Rosenbloom KR, Trumbower H, Weirauch M, Cooper DN, Stenson PD, Ma B, Brent M, Arumugam M, Shteynberg D, Copley RR, Taylor MS, Riethman H, Mudunuri U, Peterson J, Guyer M, Felsenfeld A, Old S, Mockrin S, Collins F; Rat Genome Sequencing Project Consortium (2004) Genome sequence of the Brown Norway rat yields insights into mammalian evolution. Nature 428:493–521

Gilman SR, Iossifov I, Levy D, Ronemus M, Wigler M, Vitkup D (2011) Rare de novo variants associated with autism implicate a large functional network of genes involved in formation and function of synapses. Neuron 70:898–907

Giudice A, Trounson A (2008) Genetic modification of human embryonic stem cells for derivation of target cells. Cell Stem Cell 2:422–433

Grskovic M, Javaherian A, Strulovici B, Daley GQ (2011) Induced pluripotent stem cells-opportunities for disease modelling and drug discovery. Nat Rev Drug Discov 10:915–929

Gutierrez RC, Hung J, Zhang Y, Kertesz AC, Espina FJ, Colicos MA (2009) Altered synchrony and connectivity in neuronal networks expressing an autism-related mutation of neuroligin 3. Neuroscience 162:208–221

Ho R, Chronis C, Plath K (2011) Mechanistic insights into reprogramming to induced pluripotency. J Cell Physiol 226:868–878

Hu BY, Weick JP, Yu J, Ma LX, Zhang XQ, Thomson JA, Zhang SC (2010) Neural differentiation of human induced pluripotent stem cells follows developmental principles but with variable potency. Proc Natl Acad Sci USA 107:4335–4340

Ji W, Foo JN, O'Roak BJ, Zhao H, Larson MG, Simon DB, Newton-Cheh C, State MW, Levy D, Lifton RP (2008) Rare independent mutations in renal salt handling genes contribute to blood pressure variation. Nat Genet 40:592–599

Kim KY, Hysolli E, Park IH (2011) Neuronal maturation defect in induced pluripotent stem cells from patients with Rett syndrome. Proc Natl Acad Sci USA 108:14169–14174

Lander ES, Linton LM, Birren B, Nusbaum C, Zody MC, Baldwin J, Devon K, Dewar K, Doyle M, FitzHugh W, Funke R, Gage D, Harris K, Heaford A, Howland J, Kann L, Lehoczky J, LeVine R, McEwan P, McKernan K, Meldrim J, Mesirov JP, Miranda C, Morris W, Naylor J, Raymond C, Rosetti M, Santos R, Sheridan A, Sougnez C, Stange-Thomann N, Stojanovic N, Subramanian A, Wyman D, Rogers J, Sulston J, Ainscough R, Beck S, Bentley D, Burton J, Clee C, Carter N, Coulson A, Deadman R, Deloukas P, Dunham A, Dunham I, Durbin R, French L, Grafham D, Gregory S, Hubbard T, Humphray S, Hunt A, Jones M, Lloyd C, McMurray A, Matthews L, Mercer S, Milne S, Mullikin JC, Mungall A, Plumb R, Ross M, Shownkeen R, Sims S, Waterston RH, Wilson RK, Hillier LW, McPherson JD, Marra MA, Mardis ER, Fulton LA, Chinwalla AT, Pepin KH, Gish WR, Chissoe SL, Wendl MC, Delehaunty KD, Miner TL, Delehaunty A, Kramer JB, Cook LL, Fulton RS, Johnson DL, Minx PJ, Clifton SW, Hawkins T, Branscomb E, Predki P, Richardson P, Wenning S, Slezak T, Doggett N, Cheng JF, Olsen A, Lucas S, Elkin C, Uberbacher E, Frazier M, Gibbs RA, Muzny DM, Scherer SE, Bouck JB, Sodergren EJ, Worley KC, Rives CM, Gorrell JH, Metzker ML, Naylor SL, Kucherlapati RS, Nelson DL, Weinstock GM, Sakaki Y, Fujiyama A, Hattori M, Yada T, Toyoda A, Itoh T, Kawagoe C, Watanabe H, Totoki Y, Taylor T, Weissenbach J, Heilig R, Saurin W, Artiguenave F, Brottier P, Bruls T, Pelletier E, Robert C, Wincker P, Smith DR, Doucette-Stamm L, Rubenfield M, Weinstock K, Lee HM, Dubois J, Rosenthal A, Platzer M, Nyakatura G, Taudien S, Rump A, Yang H, Yu J, Wang J, Huang G, Gu J, Hood L, Rowen L, Madan A, Qin S, Davis RW, Federspiel NA, Abola AP, Proctor MJ, Myers RM, Schmutz J, Dickson M, Grimwood J, Cox DR, Olson MV, Kaul R, Raymond C, Shimizu N, Kawasaki K, Minoshima S, Evans GA, Athanasiou M, Schultz R, Roe BA, Chen F, Pan H, Ramser J, Lehrach H, Reinhardt R, McCombie WR, de la Bastide M, Dedhia N, Blöcker H, Hornischer K, Nordsiek G, Agarwala R, Aravind L, Bailey JA, Bateman A, Batzoglou S, Birney E, Bork P, Brown DG, Burge CB, Cerutti L, Chen HC, Church D, Clamp M, Copley RR, Doerks T, Eddy SR, Eichler EE, Furey S, Galagan J, Gilbert JG, Harmon C, Hayashizaki Y, Haussler D, Hermjakob H, Hokamp K, Jang W, Johnson LS, Jones TA, Kasif S, Kaspryzk A, KennedyS, Kent WJ, Kitts P, Koonin EV, Korf I, Kulp D, Lancet D, Lowe TM, McLysaght A, Mikkelsen T, Moran JV, Mulder N, Pollara VJ, Ponting CP, Schuler G, Schultz J, Slater G, Smit AF, Stupka E, Szustakowski J, Thierry-Mieg D, Thierry-Mieg J, Wagner L, Wallis J, Wheeler R, Williams A, Wolf YI, Wolfe KH, Yang SP, Yeh RF, Collins F, Guyer MS, Peterson J, Felsenfeld A, Wetterstrand KA, Patrinos A, Morgan MJ, de Jong P, Catanese JJ, Osoegawa

K, Shizuya H, Choi S, Chen YJ (2001) Initial sequencing and analysis of the human genome. Nature 409:860–921

Laurent LC, Ulitsky I, Slavin I, Tran H, Schork A, Morey R, Lynch C, Harness JV, Lee S, Barrero MJ, Ku S, Martynova M, Semechkin R, Galat V, Gottesfeld J, Izpisua Belmonte JC, Murry C, Keirstead HS, Park HS, Schmidt U, Laslett AL, Muller FJ, Nievergelt CM, Shamir R, Loring JF (2011) Dynamic changes in the copy number of pluripotency and cell proliferation genes in human ESCs and iPSCs during reprogramming and time in culture. Cell Stem Cell 8:106–118

Levy D, Ronemus M, Yamrom B, Lee YH, Leotta A, Kendall J, Marks S, Lakshmi B, Pai D, Ye K, Buja A, Krieger A, Yoon S, Troge J, Rodgers L, Iossifov I, Wigler M (2011) Rare de novo and transmitted copy-number variation in autistic spectrum disorders. Neuron 70:886–897

Lister R, Pelizzola M, Kida YS, Hawkins RD, Nery JR, Hon G, Antosiewicz-Bourget J, O'Malley R, Castanon R, Klugman S, Downes M, Yu R, Stewart R, Ren B, Thomson JA, Evans RM, Ecker JR (2011) Hotspots of aberrant epigenomic reprogramming in human induced pluripotent stem cells. Nature 471:68–73

Liu GH, Suzuki K, Qu J, Sancho-Martinez I, Yi F, Li M, Kumar S, Nivet E, Kim J, Soligalla RD, Dubova I, Goebl A, Plongthongkum N, Fung HL, Zhang K, Loring JF, Laurent LC, Izpisua Belmonte JC (2011) Targeted gene correction of laminopathy-associated LMNA mutations in patient-specific iPSCs. Cell Stem Cell 8:688–694

Marchetto MC, Yeo GW, Kainohana O, Marsala M, Gage FH, Muotri AR (2009) Transcriptional signature and memory retention of human-induced pluripotent stem cells. PLoS One 4:e7076

Marchetto MC, Carromeu C, Acab A, Yu D, Yeo GW, Mu Y, Chen G, Gage FH, Muotri AR (2010a) A model for neural development and treatment of Rett syndrome using human induced pluripotent stem cells. Cell 143:527–539

Marchetto MC, Winner B, Gage FH (2010b) Pluripotent stem cells in neurodegenerative and neurodevelopmental diseases. Hum Mol Genet 19:R71–R76

Marchetto MC, Brennand KJ, Boyer LF, Gage FH (2011) Induced pluripotent stem cells (iPSCs) and neurological disease modeling: progress and promises. Hum Mol Genet 20:R109–R115

Mateizel I, De Temmerman N, Ullmann U, Cauffman G, Sermon K, Van de Velde H, De Rycke M, Degreef E, Devroey P, Liebaers I, Van Steirteghem A (2006) Derivation of human embryonic stem cell lines from embryos obtained after IVF and after PGD for monogenic disorders. Hum Reprod 21:503–511

Mayshar Y, Ben-David U, Lavon N, Biancotti JC, Yakir B, Clark AT, Plath K, Lowry WE, Benvenisty N (2010) Identification and classification of chromosomal aberrations in human induced pluripotent stem cells. Cell Stem Cell 7:521–531

Muotri AR (2009) Modeling epilepsy with pluripotent human cells. Epilepsy Behav 14(Suppl 1):81–85

Muotri AR, Gage FH (2006) Generation of neuronal variability and complexity. Nature 441:1087–1093

Muotri AR, Chu VT, Marchetto MC, Deng W, Moran JV, Gage FH (2005a) Somatic mosaicism in neuronal precursor cells mediated by L1 retrotransposition. Nature 435:903–910

Muotri AR, Nakashima K, Toni N, Sandler VM, Gage FH (2005b) Development of functional human embryonic stem cell-derived neurons in mouse brain. Proc Natl Acad Sci USA 102:18644–18648

Muotri AR, Marchetto MC, Coufal NG, Oefner R, Yeo G, Nakashima K, Gage FH (2010) L1 retrotransposition in neurons is modulated by MeCP2. Nature 468:443–446

Pang ZP, Yang N, Vierbuchen T, Ostermeier A, Fuentes DR, Yang TQ, Citri A, Sebastiano V, Marro S, Sudhof TC, Wernig M (2011) Induction of human neuronal cells by defined transcription factors. Nature 476:220–223

Pasca SP, Portmann T, Voineagu I, Yazawa M, Shcheglovitov A, Pasca AM, Cord B, Palmer TD, Chikahisa S, Nishino S, Bernstein JA, Hallmayer J, Geschwind DH, Dolmetsch RE (2011) Using iPSC-derived neurons to uncover cellular phenotypes associated with Timothy syndrome. Nat Med 17:1657–1662

Pickering SJ, Minger SL, Patel M, Taylor H, Black C, Burns CJ, Ekonomou A, Braude PR (2005) Generation of a human embryonic stem cell line encoding the cystic fibrosis mutation deltaF508, using preimplantation genetic diagnosis. Reprod Biomed Online 10:390–397

Robinton DA, Daley GQ (2012) The promise of induced pluripotent stem cells in research and therapy. Nature 481:295–305

Sanders SJ, Ercan-Sencicek AG, Hus V, Luo R, Murtha MT, Moreno-De-Luca D, Chu SH, Moreau MP, Gupta AR, Thomson SA, Mason CE, Bilguvar K, Celestino-Soper PB, hoi M, Crawford EL, Davis L, Wright NR, Dhodapkar RM, DiCola M, DiLullo NM, Fernandez TV, Fielding-Singh V, Fishman DO, Frahm S, Garagaloyan R, Goh GS, Kammela S, Klei L, Lowe JK, Lund SC, McGrew AD, Meyer KA, Moffat WJ, Murdoch JD, O'Roak BJ, Ober GT, Pottenger RS, Raubeson MJ, Song Y, Wang Q, Yaspan BL, Yu TW, Yurkiewicz IR, Beaudet AL, Cantor RM, Curland M, Grice DE, Günel M, Lifton P, Mane SM, Martin DM, Shaw CA, Sheldon M, Tischfield JA, Walsh CA, Morrow EM, Ledbetter DH, Fombonne E, Lord C, Martin CL, Brooks AI, Sutcliffe JS, Cook EH Jr, Geschwind D, Roeder K, Devlin B, State MW (2011) Multiple recurrent de novo CNVs, including duplications of the 7q1123 Williams syndrome region, are strongly associated with autism. Neuron 70:863–885

Saporta MA, Grskovic M, Dimos JT (2011) Induced pluripotent stem cells in the study of neurological diseases. Stem Cell Res Ther 2:37

Scheuner D, Eckman C, Jensen M, Song X, Citron M, Suzuki N, Bird TD, Hardy J, Hutton M, Kukull W, Larson E, Levy-Lahad E, Viitanen M, Peskind E, Poorkaj P, Schellenberg G, Tanzi R, Wasco W, Lannfelt L, Selkoe D, Younkin S (1996) Secreted amyloid beta-protein similar to that in the senile plaques of Alzheimer's disease is increased in vivo by the presenilin 1 and 2 and APP mutations linked to familial Alzheimer's disease. Nat Med 2:864–870

Silverman JL, Yang M, Lord C, Crawley JN (2010) Behavioural phenotyping assays for mouse models of autism. Nat Rev Neurosci 11:490–502

Soldner F, Laganiere J, Cheng AW, Hockemeyer D, Gao Q, Alagappan R, Khurana V, Golbe LI, Myers RH, Lindquist S, Zhang L, Guschin D, Fong LK, Vu BJ, Meng X, Urnov FD, Rebar EJ, Gregory PD, Zhang HS, Jaenisch R (2011) Generation of isogenic pluripotent stem cells differing exclusively at two early onset Parkinson point mutations. Cell 146:318–331

Splawski I, Timothy KW, Sharpe LM, Decher N, Kumar P, Bloise R, Napolitano C, Schwartz PJ, Joseph RM, Condouris K, Tager-Flusberg H, Priori SG, Sanguinetti MC, Keating MT (2004) Ca(V)12 calcium channel dysfunction causes a multisystem disorder including arrhythmia and autism. Cell 119:19–31

Splawski I, Yoo DS, Stotz SC, Cherry A, Clapham DE, Keating MT (2006) CACNA1H mutations in autism spectrum disorders. J Biol Chem 281:22085–22091

State MW, Levitt P (2011) The conundrums of understanding genetic risks for autism spectrum disorders. Nat Neurosci 14:1499–1506

Takahashi K, Yamanaka S (2006) Induction of pluripotent stem cells from mouse embryonic and adult fibroblast cultures by defined factors. Cell 126:663–676

Takahashi K, Tanabe K, Ohnuki M, Narita M, Ichisaka T, Tomoda K, Yamanaka S (2007) Induction of pluripotent stem cells from adult human fibroblasts by defined factors. Cell 131:861–872

Thomson JA, Itskovitz-Eldor J, Shapiro SS, Waknitz MA, Swiergiel JJ, Marshall VS, Jones JM (1998) Embryonic stem cell lines derived from human blastocysts. Science 282:1145–1147

Tropea D, Giacometti E, Wilson NR, Beard C, McCurry C, Fu DD, Flannery R, Jaenisch R, Sur M (2009) Partial reversal of Rett Syndrome-like symptoms in MeCP2 mutant mice. Proc Natl Acad Sci USA 106:2029–2034

Urbach A, Schuldiner M, Benvenisty N (2004) Modeling for Lesch–Nyhan disease by gene targeting in human embryonic stem cells. Stem Cells 22:635–641

Urbach A, Bar-Nur O, Daley GQ, Benvenisty N (2010) Differential modeling of fragile X syndrome by human embryonic stem cells and induced pluripotent stem cells. Cell Stem Cell 6:407–411

Verlinsky Y, Strelchenko N, Kukharenko V, Rechitsky S, Verlinsky O, Galat V, Kuliev A (2005) Human embryonic stem cell lines with genetic disorders. Reprod Biomed Online 10:105–110

Vierbuchen T, Wernig M (2011) Direct lineage conversions: unnatural but useful? Nat Biotechnol 29:892–907

Vierbuchen T, Ostermeier A, Pang ZP, Kokubu Y, Sudhof TC, Wernig M (2011) Direct conversion of fibroblasts to functional neurons by defined factors. Nature 463:1035–1041

Waterston RH, Lindblad-Toh K, Birney E, Rogers J, Abril JF, Agarwal P, Agarwala R, Ainscough R, Alexandersson M, An P, Antonarakis SE, Attwood J, Baertsch R, Bailey J, Barlow K, Beck S, Berry E, Birren B, Bloom T, Bork P, Botcherby M, Bray N, Brent MR, Brown DG, Brown SD, Bult C, Burton J, Butler J, Campbell RD, Carninci P, Cawley S, Chiaromonte F, Chinwalla AT, Church DM, Clamp M, Clee C, Collins FS, Cook LL, Copley RR, Coulson A, Couronne O, Cuff J, Curwen V, Cutts T, Daly M, David R, Davies J, Delehaunty KD, Deri J, Dermitzakis ET, Dewey C, Dickens NJ, Diekhans M, Dodge S, Dubchak I, Dunn DM, Eddy SR, Elnitski L, Emes RD, Eswara P, Eyras E, Felsenfeld A, Fewell GA, Flicek P, Foley K, Frankel WN, Fulton LA, Fulton RS, Furey TS, Gage D, Gibbs RA, Glusman G, Gnerre S, Goldman N, Goodstadt L, Grafham D, Graves TA, Green ED, Gregory S, Guigó R, Guyer M, Hardison RC, Haussler D, Hayashizaki Y, Hillier LW, Hinrichs A, Hlavina W, Holzer T, Hsu F, Hua A, Hubbard T, Hunt A, Jackson I, Jaffe DB, Johnson LS, Jones M, Jones TA, Joy A, Kamal M, Karlsson EK, Karolchik D, Kasprzyk A, Kawai J, Keibler E, Kells C, Kent WJ, Kirby A, Kolbe DL, Korf I, Kucherlapati RS, Kulbokas EJ, Kulp D, Landers T, Leger JP, Leonard S, Letunic I, Levine R, Li J, Li M, Lloyd C, Lucas S, Ma B, Maglott DR, Mardis ER, Matthews L, Mauceli E, Mayer JH, McCarthy M, McCombie WR, McLaren S, McLay K, McPherson JD, Meldrim J, Meredith B, Mesirov JP, Miller W, Miner TL, Mongin E, Montgomery KT, Morgan M, Mott R, Mullikin JC, Muzny DM, Nash WE, Nelson JO, Nhan MN, Nicol R, Ning Z, Nusbaum C, O'Connor MJ, Okazaki Y, Oliver K, Overton-Larty E, Pachter L, Parra G, Pepin KH, Peterson J, Pevzner P, Plumb R, Pohl CS, Poliakov A, Ponce TC, Ponting CP, Potter S, Quail M, Reymond A, Roe BA, Roskin KM, Rubin EM, Rust AG, Santos R, Sapojnikov V, Schultz B, Schultz J, Schwartz MS, Schwartz S, Scott C, Seaman S, Searle S, Sharpe T, Sheridan A, Shownkeen R, Sims S, Singer JB, Slater G, Smit A, Smith DR, Spencer B, Stabenau A, Stange-Thomann N, Sugnet C, Suyama M, Tesler G, Thompson J, Torrents D, Trevaskis E, Tromp J, Ucla C, Ureta-Vidal A, Vinson JP, Von Niederhausern AC, Wade CM, Wall M, Weber RJ, Weiss RB, Wendl MC, West AP, Wetterstrand K, Wheeler R, Whelan S, Wierzbowski J, Willey D, Williams S, Wilson RK, Winter E, Worley KC, Wyman D, Yang S, Yang SP, Zdobnov EM, Zody MC, Lander ES Mouse Genome Sequencing Consortium (2002) Initial sequencing and comparative analysis of the mouse genome. Nature 420:520–562

Weiss LA, Shen Y, Korn JM, Arking DE, Miller DT, Fossdal R, Saemundsen E, Stefansson H, Ferreira MA, Green T, Platt OS, Ruderfer DM, Walsh CA, Altshuler D, Chakravarti A, Tanzi RE, Stefansson K, Santangelo SL, Gusella JF, Sklar P, Wu BL, Daly MJ; Autism Consortium (2008) Association between microdeletion and microduplication at 16p112 and autism. N Engl J Med 358:667–675

Yu J, Vodyanik MA, Smuga-Otto K, Antosiewicz-Bourget J, Frane JL, Tian S, Nie J, Jonsdottir GA, Ruotti V, Stewart R, Slukvin II, Thomson JA (2007) Induced pluripotent stem cell lines derived from human somatic cells. Science 318:1917–1920

On the Search for Reliable Human Aging Models: Understanding Aging by Nuclear Reprogramming

Ignacio Sancho-Martinez, Emmanuel Nivet, and Juan Carlos Izpisua Belmonte

Abstract Reprogramming technologies, and particularly the generation of induced pluripotent stem cells (iPSCs), have demonstrated the possibility of personalized disease modeling in a dish. Importantly, the fact that pluripotent stem cells can give rise to all cell types of an organism, along with the technical progress allowing for their isolation, brings to mind fantasies like the fountain of youth and eternal regeneration and represents one of the most promising scientific fields with clinical implications. Furthermore, increasing evidence indicates that aging "defects" observed in patient somatic cells could be erased or alleviated by direct reprogramming towards pluripotency and rapidly recapitulated upon directed differentiation to specific cell lineages (Liu et al., Nature 472:221–225, 2011a). Thus, iPSC models of aging facilitate human aging studies by shortening the time required for physiological manifestation of aging-related defects from years, in the case of a human being, to days when stem cell models are applied. Moreover, the combination of gene-editing and iPSC models of aging will also allow for the generation of precisely targeted reporter cell lines of high value for studying normal differentiation processes and high throughput screens. However, a major concern regarding the use of iPSCs for disease modeling has to be taken into account prior to their broad application in drug discovery studies, which is that the use of patient-derived iPSCs bears another important experimental limitation, the lack of appropriate genetically matched control lines (Soldner et al., Cell 146:318–331, 2011; Liu et al., Cell Stem Cell 8:688–694, 2011b).

I. Sancho-Martinez • E. Nivet
Gene Expression Laboratory, Salk Institute for Biological Studies, 10010 North Torrey Pines Road, La Jolla, CA 92037, USA

J.C.I. Belmonte (✉)
Gene Expression Laboratory, Salk Institute for Biological Studies, 10010 North Torrey Pines Road, La Jolla, CA 92037, USA

Center for Regenerative Medicine in Barcelona, Dr. Aiguader 88, Barcelona 08003, Spain
e-mail: belmonte@salk.edu; izpisua@cmrb.eu

F.H. Gage and Y. Christen (eds.), *Programmed Cells from Basic Neuroscience to Therapy*, Research and Perspectives in Neurosciences 20,
DOI 10.1007/978-3-642-36648-2_11, © Springer-Verlag Berlin Heidelberg 2013

In this chapter we will discuss the most recent advancements in the use of pluripotent stem cells as models of disease with special emphasis on their use, alongside gene editing, for the study of human aging and its associated pathologies.

Introduction

Due to increased life span and fertility, the world population of those over 60 years of age is expected to increase to more than 2.4 billion (21.8 % of the total population) in 2050 (Lutz et al. 2008). Accompanying the aging population is the advent of aging-associated diseases affecting a large number of elderly people, such as cardiovascular disease, diabetes, cancer, and various neurodegenerative disorders (Alwan et al. 2010). The increasing aging population and numerous aging-associated diseases are of public concern and are thus spurring the need for research in this area. Life span and aging-related physiological decline, or degeneration, have been extensively studied in model organisms, dating back to 1935 with McCay and colleagues, who noted that caloric restriction extended the life span of rodents. A considerable body of data in *C. elegans*, yeast, *Drosophila*, and mice has started to unravel the mechanism of aging and aging-related tissue degeneration (Vig and Campisi 2008; Panowski et al. 2007). However, to date, there has been a lack of comprehensive approaches and reliable experimental models suitable for human aging research and anti-aging therapeutic development.

Human Aging Studies: Reality or Utopia?

Progeroid syndromes, which share many features with normal human aging, have been traditionally studied in animal models that, even though informative, do not fully resemble the human system (Chen et al. 2012). These syndromes include Hutchinson–Gilford progeria syndrome (HGPS, also known as 'progeria of the child'), Werner syndrome (WS, also known as 'progeria of the adult'), Cockayne syndrome, and Klinefelter and Turner syndromes (Jin 2010; Partridge et al. 2011; Scaffidi et al. 2006; Yang et al. 2006). Abnormality in the nuclear lamina and imperfection in DNA repair systems are two major causes of accelerated human aging syndromes that are present in HGPS and WS, respectively (Jin 2010; Partridge et al. 2011; Scaffidi et al. 2006; Yang et al. 2006; Itahana et al. 2004; Jucker 2010). Indeed, nuclear architecture has been linked to a number of cellular processes, including epigenetic modifications and gene expression. As such, defects in the nuclear envelope machinery have been demonstrated to correlate with the manifestation of a number of diseases as well as aging. Recently, a number of studies have implicated lamin proteins, critical components of the nuclear envelope, in accelerated senescence leading to the development of the progeria syndrome and to premature aging disease (Li et al. 2011; Liu et al. 2011a; Scaffidi et al. 2006; Chen et al. 2012; Campisi and d'Adda di Fagagna 2007). Furthermore, mutations in Lamin-A, and accordingly in the aberrant nuclear envelope, can result in a number of other diseases generally referred to as laminopathies. Lamin-related nuclear defects directly correlate with senescence, thus making the disease phenotype more implicated in physiological aging. Accordingly, it is probable that diseases whose manifestation occurs during the late stages of life might be connected to aberrant nuclear architecture as well as progressive nuclear envelope destruction.

HGPS patients usually die at a median age of 13 due to myocardial infarction and stroke. These patients commonly show growth retardation after 1 year of age, followed by aged facies, sclerotic skin, decreased joint mobility, early hair loss, and cardiovascular problems. Arteriosclerosis and premature loss of vascular smooth muscle cells (SMCs) are common characteristics of this syndrome as well as in elderly people. The genetic basis of HGPS was not uncovered until 2003, when a single nucleotide substitution of LMNA, whose encoding products are A-type nuclear lamins, lamins A and C, were found to be the offenders in most instances (Jin 2010; Partridge et al. 2011; Scaffidi et al. 2006; Yang et al. 2006). The prevalent G608G LMNA mutation activates a cryptic splicing site in pre-lamin A, leading to a truncated mutant of lamin A known as progerin. Progerin accumulation results in abnormal nuclear envelopes, misregulation of the heterochromatin and nuclear lamina proteins, and many other nuclear defects, including attrited telomeres and genomic instability. The pathogenic progerin is mainly present in vascular SMCs, Mesenchymal stem cells (MSCs), dermal fibroblasts, and keratinocytes. It has been recently reported that progerin and telomere dysfunction collaborate to trigger human fibroblast senescence, shedding light on the question of how progerin participates in the normal aging process (Scaffidi et al. 2006). Unfortunately, traditional models normally focus on end-point primary human cells and/or animal models that, even though phenocopying human HGPS, do not share the same molecular triggers observed in human patients; neither are they able to recapitulate development of the disease (Murga et al. 2009; Hinkal et al. 2009; Yang et al. 2006; Li et al. 2011; Jucker 2010).

As previously discussed, striking similarities have been reported between physiological aging and the premature aging disease HGPS. HGPS is caused by constitutive production of progerin, a mutant form of the nuclear architectural protein lamin A. Progerin acts in a dominant gain-of-function fashion by accumulating at the nuclear periphery and altering nuclear lamina structure (Jin 2010; Partridge et al. 2011; Scaffidi et al. 2006; Yang et al. 2006). Cells from HGPS patients exhibit extensive nuclear defects, including abnormal chromatin structure, increased DNA damage, and shortened telomeres (Cao et al. 2011), and all these factors have been linked to physiological cell senescence. The common causes of death in HGPS patients are chronic conditions commonly found in elderly people, like coronary artery disease and stroke, conditions that probably are a consequence of the vascular, heart, fat, and bone abnormalities found in HGPS patients. In recent years, a number of animal models have served to advance our understanding of

HGPS, yet such models phenocopy HGPS symptoms rather than model the actual molecular implications of the disease. For example, whereas Lmna−/− and LmnaD9 mice, expressing a deleted form of Lmna (deleted for exon 9 with the in-frame removal of 40 amino acids of lamin A/C), have been developed, there are distinct differences between Lmna−/−, LmnaD9, and human HGPS (Misteli 2011; Scaffidi et al. 2008; Murga et al. 2009; Hinkal et al. 2009; Yang et al. 2006; Li et al. 2011; Jucker 2010). The Lmna−/− mouse does not express a full-length lamin A protein, whereas the LmnaD9 mouse expresses a farnesylated lamin A-DExon9 mutant protein that, even though similar, is not identical to the heterozygous expression of the mutant protein in HGPS. Thus, and even though progress has been made, Progeria still lacks reliable human models for the study of the molecular implications, causes and consequences of this premature aging syndrome. An alternative could be the use of stem cell models recapitulating physiological manifestation and progression of aging-associated phenotypes.

Stem Cell Models of Disease: An Alternative to Animal Studies on Aging

Until recently, one of the major problems related to the use of human pluripotent stem cells was more societal than technical. The need for embryonic material, and its consequent destruction for the isolation of embryonic stem cells (ESCs), has been the source of disputes in the political and religious arenas, as well as the media and the scientific community itself. Similarly, the ability to generate pluripotent cells by using Somatic Cell Nuclear Transfer (SCNT) technology has led to the fear of the possibility of therapeutic cloning, in which human clones could eventually be utilized to cure the "original" individual. Therefore, strict regulatory laws regarding the use of embryonic- and SCNT-derived ESCs have been established in a number of countries. In 2006, Takahashi and Yamanaka revealed to the world the ability to experimentally generate mouse and later human (Takahashi et al. 2006, 2007), pluripotent stem cells without the need for embryonic material. Using a funnel strategy, they overexpressed in adult mouse fibroblasts 24 genes previously identified as playing pivotal roles in the maintenance of pluripotency in ESCs. By exploring multiple combinations and with a reductionist approach, they ended up with the identification of four transcription factors, namely Oct3/4, Sox2, c-Myc and Klf4, allowing the reversion of adult fibroblasts back to an ES-like cell phenotype when maintained in pluripotent culture conditions previously established. This major breakthrough generated incredible excitement in the scientific community. Later on, the Thomson laboratory reported the possibility of reprogramming somatic cells by replacing two of the "Yamanaka factors," c-Myc and Klf4, with Nanog and Lin28 (Yu et al. 2007). Later on, several laboratories reported that pluripotency could be achieved by using three, two and even only one of these factors, depending on the somatic cell type that was started with.

The observation that Oct4 alone is able to revert neural stem cells back to an ES-like cell (Kim et al. 2009), has revealed Oct4 as the core transcription factor for pluripotency. Since the initial reports describing iPSC generation by viral-based/integrative approaches, including the use of mono- and poly-cistronic vectors, laboratories worldwide rapidly raced to develop alternative technologies to avoid the integration of exogenous DNA into the host genome. To date, several integration-free reprogramming methods have been described, including (1) Cre-recombinase excisable viruses, (2) non-integrating adenoviruses, (3) expression plasmids, (4) piggyBac transposition, (5) episomal vectors, (6) delivery of reprogramming proteins, and (7) delivery of mRNAs (Sancho-Martinez et al. 2011). More recently, the possibility of reprogramming somatic cells by overexpression of a specific cluster of miRNAs (miR-302/367) was reported, raising the possibility of rapidly developing another integration-free reprogramming method (Anokye-Danso et al. 2011; Miyoshi et al. 2011).

Regardless of the methodology used for their derivation, pluripotent stem cells, due to their ability to virtually generate all cell types composing an adult organism, have offered the possibility to study human developmental biology in a culture dish. Reciprocally, knowledge acquired from developmental studies in other model systems has facilitated our understanding of lineage commitment. Beyond the use of pluripotent cells for revealing developmental processes, the possibility of reproducibly driving the differentiation of pluripotent stem cells towards a specific cell population represents a major hope for curing many diseases whose origins are in a cellular deficiency or malfunction and for which no efficient molecules have been found yet. As we will further discuss, the differentiation of pluripotent stem cells into a clinically relevant cell type, as well as the possibility for gene correction of mutant genes, represents an alternative for personalized cell therapy and also for the development of platforms useful for drug-screening and disease modeling of special relevance for those diseases and/or physiological manifestations for which reliable animal models are not available, such as human aging (Tiscornia et al. 2011; Liu et al. 2011a).

Fine-Tuning Cellular Models; Gene Editing and Stem Cells

Of particular interest, the development of gene-editing technologies in combination with the generation of patient-specific iPSC could represent a merge of both stem cell and traditional gene therapy and emerge as one of the most reliable models of human aging and disease in general. Patient-derived iPSCs bearing monogenic mutations responsible for disease development are suitable material for in vitro correction of the mutant gene and further re-transplantation of the corrected cells into the patient. Moreover, gene-targeting technologies in mouse ESCs have made enormous contributions to the understanding of gene function, animal development and disease pathologies (Liu et al. 2012).

However, translating the success of gene targeting in mouse ESCs into human ESCs or iPSCs has been challenging. Random integration of transgenes, a common feature with most of the traditional reprogramming approaches, has been the predominant method for modifying the human genome in the past. Indeed, the drawbacks of this approach are increasingly recognized, including the potential for insertional mutagenesis leading to tumor formation. Although random integration of transgenes mediated by viral transduction or transposable elements still holds its value in many applications, thanks to its simplicity and effectiveness, the field has moved beyond it and is in need of more precise ways to modify the human genome. Indeed, a number of recent publications have reported the successful correction of genes bearing mutations responsible for disease (Li et al. 2011; Liu et al. 2011b; Soldner et al. 2011; Howden et al. 2011; Papapetrou et al. 2011). Thus, different technologies can be applied for the genetic restoration of wild-type gene copies, which, in the context of a monogenic disease, ultimately leads to phenotypic recovery and function improvement. Thus, gene-correction, as opposed to traditional gene therapy, in which genetic complementation rather than actual correction of the mutant gene is exploited, leads to the splicing of the mutant gene and its replacement by homologous recombination-mediated insertion of the wild-type version of the gene.

So far, a number of different technologies have been developed, each with hallmark advantages and disadvantages. Yet, the most common issues associated with gene targeting can be generally summarized as follows: (1) low efficiencies, particularly in pluripotent cells and transcriptionally inactive loci; and (2) off target effects manifested by high toxicity, high incidence of random integration and other mutagenic responses due to the targeting process itself. Perhaps one of the most extended technologies nowadays involves the use of Zinc Finger Nuclease (ZFN; Hockemeyer et al. 2009; Soldner et al. 2011). Briefly, ZFNs are engineered proteins that recognize specific sequences of the genome. Nuclease activity leads to the generation of double strand breaks (DSBs) on the DNA that can be further repaired by two different endogenous mechanisms: homologous recombination, leading to the successful correction of the targeted gene, and non-homologous end-joining, an error-prone DNA-repair mechanism leading to the generation of mutations in the host genome. Several other technologies have been reported so far, including the use of Helper Dependent Adenoviruses (HDAdV; Suzuki et al. 2008; Liu et al. 2011b), bacterial artificial chromosomes (BAC; Yang and Seed 2003), and Transcription Activator-Like Effector Nucleases (TALENs; Cermak et al. 2011), among others, and represent some of the most promising approaches for efficient gene-editing. Of note is the fact that the broad applications of gene-editing technologies cover not only gene correction but also the generation of precisely targeted reporter cell lines that are extremely valuable for studying normal differentiation processes and high throughput screens. Moreover, specific knock-in as well as knock-out of genes of interest in pluripotent cell lines represent unmatched tools for molecular studies. In this regard, patient-derived iPSCs not only hold immense promise in terms of gene-correction and regenerative medicine but also allow for concise analysis of the molecular mechanisms leading to manifestation and progression of a specific disease.

Disease modeling subsequently presents the possibility of in vitro drug discovery, testing and development of personalized therapies (Dimos et al. 2008; Marchetto et al. 2010; Liu et al. 2011a; Brennand et al. 2011). However, three major concerns regarding the use of iPSCs for disease modeling have to be taken into account prior to their broad application in drug discovery studies. Firstly, considering the accumulation of genetic and epigenetic abnormalities during the reprogramming process, it is thought that patient-derived iPSCs might present abnormal functionality. In such a situation, genetic abnormalities leading to the development of disease might contribute and/or synergize to the defective reprogramming of somatic cells into iPSCs (a question not yet addressed), which might lead to the wrong interpretation of results. Secondly, two recent reports have pointed out important differences between ESCs and iPSCs in specific disease contexts (Urbach et al. 2010). Thirdly, the use of patient-derived iPSCs bears, by definition, another important experimental limitation, the lack of appropriate control lines. Of relevance is the fact that reprogramming-associated genetic and epigenetic defects, even though clustering in specific cancer-related pathways, seem to be random in terms of specific genes. Thus, the use of different iPSC lines – ones derived from a diseased patient and ones derived from healthy individuals – might indeed bear a number of epigenetic and genetic differences, leading to the wrong interpretation of the results during drug discovery and disease modeling studies. In such a case, a novel approach could take advantage of the development of novel gene-editing technologies in two different ways: (1) through the generation of isogenic, genetically matched iPSC lines; and (2) through the manipulation of "real" pluripotent cells, ESCs, and allowing not for gene correction but for the modification of the cells in the opposite direction, i.e., the generation of disease-specific ESCs (Soldner et al. 2011). In such a paradigm, generation of ESCs bearing mutant genes responsible for disease might well represent a more reliable source of pluripotent cells to model disease, as direct splicing of wild-type genes and knock-in of mutant genes in already pluripotent cells would bypass the reprogramming steps and all their associated "side-effects." Along this line, bypassing the reprogramming steps by generation of disease-specific ESCs might short-cut the necessity for validation of patient-derived iPSCs and the potential mis-conclusions that could arise from the use of a defective disease model in vitro. Whereas both approaches have their own advantages and disadvantages and it remains unclear which one will ultimately be the more reliable for modeling disease, the fact is that, when combined with the use of pluripotent cells, gene-correction approaches could ultimately lead to the cure or alleviation of human disease as well as represent more reliable models of disease, due to the inherent recapitulation of embryonic development once pluripotent cells are subjected to differentiation.

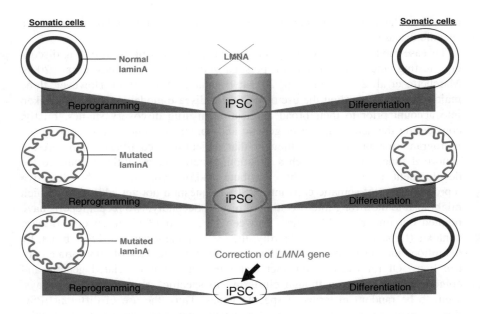

Fig. 1 Gene-correction abolishes aging manifestations in differentiated iPSCs. Reprogramming of HGPS patients into iPSCs restores all nuclear abnormalities and resets cellular defects that are recapitulated upon re-differentiation, thus establishing a reliable model of premature aging in a dish (*middle panel*). Genetic correction of LMNA, the mutant gene responsible for HGPS, restores nuclear normalcy and HGPS phenotype manifestation in iPSC-differentiated cells

iPSC Models of HGPS: Modeling Human Aging in a Dish

Recently, we have evaluated whether HGPS somatic cells could be induced to a pluripotent state and thus serve as a model of premature aging upon re-differentiation (Fig. 1). Three HGPS primary fibroblast lines, originally isolated from patients with the classical LMNA mutation (Gly608Gly), were transduced with retroviruses encoding OCT4, SOX2, KLF4, c-MYC, and GFP. Control and HGPS iPSC lines demonstrated pluripotent gene expression, demethylation of the OCT4 promoter, and transgene silencing (Liu et al. 2011a).

Interestingly, we found that the expression and localization of nuclear lamina components, heterochromatin markers, in HGPS-iPSCs were reset to a status similar to those of BJ-iPSCs (Liu et al. 2011a). Upon re-differentiation, progerin mRNA was selectively induced in differentiated HGPS-iPSCs but not in differentiated BJ-iPSCs. In contrast, lamin A was upregulated in both HGPS-iPSCs and BJ-iPSCs upon the same differentiation procedure. This reversible suppression of progerin expression by reprogramming, and subsequent reactivation upon differentiation, provides a unique model system to study human premature aging pathologies (Fig. 1; Liu et al. 2011a). We therefore next asked whether SMCs differentiated in vitro from HGPS-iPSCs exhibited premature senescence phenotypes similar to those observed in HGPS patients. Indeed, an increasing frequency of misshapen nuclei and a loss of heterochromatin mark H3K9Me3 were specifically observed in HGPS-SMCs after serial passaging. The late passages of HGPS-SMCs showed the typical characteristics of premature senescence, including increased senescence-associated-β-Gal (SA-β-Gal) staining, reduced telomere length, reduced number of Ki67-positive cells, and compromised cell proliferation. We also found a selective upregulation of senescence-related transcripts in HGPS-iPSC-derived SMCs (Liu et al. 2011a).

Upon establishment of an HGPS stem cell model, we next aimed to develop gene-editing technologies suitable for gene correction and the generation of genetically matched isogenic iPSC lines. Thus, we engineered a HDAd-based gene-correction vector (LMNA-c-HDAdV; Liu et al. 2011b). Indeed, the use of a single HDAdV was sufficient for the correction of different mutations spanning a substantially large region of the LMNA gene (Liu et al. 2011b). Targeted correction of LMNA led to the restored expression of wild-type Lamin A and the abolition of progerin expression. Consequently, lack of progerin expression resulted in the correction of the disease-associated cellular phenotypes (Fig. 1). Thus, our methodologies represent an efficient way to allow for the generation of genetically matched iPSCs, serving as a more reliable control for disease modeling and drug discovery (Liu et al. 2011b; Soldner et al. 2011; Hockemeyer et al. 2009).

Conclusion and Perspective

Altogether, the use of iPSC disease models could provide novel insights into the molecular mechanisms of human aging and also create an unprecedented platform for developing novel drugs to facilitate healthy aging and prevent or cure various aging-related diseases. More importantly, the genetic correction of monogenic mutations responsible for the development of disease and generation of isogenic iPSC/ESC lines may not only contribute towards more reliable and experimentally matched control sets in future drug discovery studies but may also provide the opportunity for combining gene therapy and regenerative medicine in the development of future therapeutic schemes targeting aging-related degenerative disorders.

Acknowledgments Work in the laboratory of J.C.I.B. was supported by grants from MINECO, Fundacion Cellex, G. Harold and Leila Y. Mathers Charitable Foundation, The Leona M. and Harry B. Helmsley Charitable Trust, The Ellison Medical Foundation and IPSEN Foundation.

References

Alwan A, Maclean DR, Riley LM, d'Espaignet ET, Mathers CD, Stevens GA, Bettcher D (2010) Monitoring and surveillance of chronic non-communicable diseases: progress and capacity in high-burden countries. Lancet 376:1861–1868

Anokye-Danso F, Trivedi CM, Juhr D, Gupta M, Cui Z, Tian Y, Zhang Y, Yang W, Gruber PJ, Epstein JA, Morrisey EE (2011) Highly efficient miRNA-mediated reprogramming of mouse and human somatic cells to pluripotency. Cell Stem Cell 8:376–388

Brennand KJ, Simone A, Jou J, Gelboin-Burkhart C, Tran N, Sangar S, Li Y, Mu Y, Chen G, Yu D, McCarthy S, Sebat J, Gage FH (2011) Modelling schizophrenia using human induced pluripotent stem cells. Nature 473:221–225

Campisi J, d'Adda di Fagagna F (2007) Cellular senescence: when bad things happen to good cells. Nat Rev Mol Cell Biol 8:729–740

Cao K, Blair CD, Faddah DA, Kieckhaefer JE, Olive M, Erdos MR, Nabel EG, Collins FS (2011) Progerin and telomere dysfunction collaborate to trigger cellular senescence in normal human fibroblasts. J Clin Invest 121:2833–2844

Cermak T, Doyle EL, Christian M, Wang L, Zhang Y, Schmidt C, Baller JA, Somia NV, Bogdanove AJ, Voytas DF (2011) Efficient design and assembly of custom TALEN and other TAL effector-based constructs for DNA targeting. Nucl Acids Res 39:e82

Chen CY, Chi YH, Mutalif RA, Starost MF, Myers TG, Anderson SA, Stewart CL, Jeang KT (2012) Accumulation of the inner nuclear envelope protein sun1 is pathogenic in progeric and dystrophic laminopathies. Cell 149:565–577

Dimos JT, Rodolfa KT, Niakan KK, Weisenthal LM, Mitsumoto H, Chung W, Croft GF, Saphier G, Leibel R, Goland R, Wichterle H, Henderson CE, Eggan K (2008) Induced pluripotent stem cells generated from patients with ALS can be differentiated into motor neurons. Science 321:1218–1221

Hinkal GW, Gatza CE, Parikh N, Donehower LA (2009) Altered senescence, apoptosis, and DNA damage response in a mutant p53 model of accelerated aging. Mech Ageing Dev 130:262–271

Hockemeyer D, Soldner F, Beard C, Gao Q, Mitalipova M, DeKelver RC, Katibah GE, Amora R, Boydston EA, Zeitler B, Meng X, Miller JC, Zhang L, Rebar EJ, Gregory PD, Urnov FD, Jaenisch R (2009) Efficient targeting of expressed and silent genes in human ESCs and iPSCs using zinc-finger nucleases. Nature Biotechnol 27:851–857

Howden SE, Gore A, Li Z, Fung HL, Nisler BS, Nie J, Chen G, McIntosh BE, Gulbranson DR, Diol NR, Taapken SM, Vereide DT, Montgomery KD, Zhang K, Gamm DM, Thomson JA (2011) Genetic correction and analysis of induced pluripotent stem cells from a patient with gyrate atrophy. Proc Natl Acad Sci USA 108(16):6537–6542

Itahana K, Campisi J, Dimri GP (2004) Mechanisms of cellular senescence in human and mouse cells. Biogerontology 5:1–10

Jin K (2010) Modern biological theories of aging. Aging Dis 1:72–74

Jucker M (2010) The benefits and limitations of animal models for translational research in neurodegenerative diseases. Nat Med 16:1210–1214

Kim JB, Sebastiano V, Wu G, Araúzo-Bravo MJ, Sasse P, Gentile L, Ko K, Ruau D, Ehrich M, van den Boom D, Meyer J, Hübner K, Bernemann C, Ortmeier C, Zenke M, Fleischmann BK, Zaehres H, Schöler HR (2009) Oct4-induced pluripotency in adult neural stem cells. Cell 136:411–419

Li H, Haurigot V, Doyon Y, Li T, Wong SY, Bhagwat AS, Malani N, Anguela XM, Sharma R, Ivanciu L, Murphy SL, Finn JD, Khazi FR, Zhou S, Paschon DE, Rebar EJ, Bushman FD, Gregory PD, Holmes MC, High KA (2011) In vivo genome editing restores haemostasis in a mouse model of haemophilia. Nature 475:217–221

Liu GH, Barkho BZ, Ruiz S, Diep D, Qu J, Yang SL, Panopoulos AD, Suzuki K, Kurian L, Walsh C, Thompson J, Boue S, Fung HL, Sancho-Martinez I, Zhang K, Yates J 3rd, Izpisua Belmonte JC (2011a) Recapitulation of premature ageing with iPSCs from Hutchinson-Gilford progeria syndrome. Nature 472:221–225

Liu GH, Suzuki K, Qu J, Sancho-Martinez I, Yi F, Li M, Kumar S, Nivet E, Kim J, Soligalla RD, Dubova I, Goebl A, Plongthongkum N, Fung HL, Zhang K, Loring JF, Laurent LC, Izpisua Belmonte JC (2011b) Targeted gene correction of laminopathy-associated LMNA mutations in patient-specific iPSCs. Cell Stem Cell 8:688–694

Liu GH, Sancho-Martinez I, Izpisua Belmonte JC (2012) Cut and paste: restoring cellular function by gene-correction. Cell Res 22:283–284

Lutz W, Sanderson W, Scherbov S (2008) The coming acceleration of global population ageing. Nature 451:716–719

Marchetto MCN, Carromeu C, Acab A, Yu D, Yeo GW, Mu Y, Chen G, Gage FH, Muotri AR (2010) A model for neural development and treatment of Rett syndrome using human induced pluripotent stem cells. Cell 143:527–539

Misteli T (2011) HGPS-derived iPSCs for the ages. Cell Stem Cell 8:4–6

Miyoshi N, Ishii H, Nagano H, Haraguchi N, Dewi DL, Kano Y, Nishikawa S, Tanemura M, Mimori K, Tanaka F, Saito T, Nishimura J, Takemasa I, Mizushima T, Ikeda M, Yamamoto H, Sekimoto M, Doki Y, Mori M (2011) Reprogramming of mouse and human cells to pluripotency using mature microRNAs. Cell Stem Cell 8:633–638

Murga M, Bunting S, Montaña MF, Soria R, Mulero F, Cañamero M, Lee Y, McKinnon PJ, Nussenzweig A, Fernandez-Capetillo O (2009) A mouse model of ATR-Seckel shows embryonic replicative stress and accelerated aging. Nat Genet 41:891–898

Panowski SH, Wolff S, Aguilaniu H, Durieux J, Dillin A (2007) PHA-4/Foxa mediates diet-restriction-induced longevity of C. elegans. Nature 447:550–555

Papapetrou EP, Lee G, Malani N, Setty M, Riviere I, Tirunagari LM, Kadota K, Roth SL, Giardina P, Viale A, Leslie C, Bushman FD, Studer L, Sadelain M (2011) Genomic safe harbors permit high β-globin transgene expression in thalassemia induced pluripotent stem cells. Nature Biotechnol 29:73–78

Partridge L, Thornton J, Bates G (2011) The new science of ageing. Philos Trans R Soc Lond B Biol Sci 366:6–8

Sancho-Martinez I, Nivet E, Izpisua Belmonte JC (2011) The labyrinth of nuclear reprogramming. J Mol Cell Biol 3:327–329

Scaffidi P, Misteli T (2006) Lamin A-dependent nuclear defects in human aging. Science 312:1059–1063

Scaffidi P, Misteli T (2008) Lamin A-dependent misregulation of adult stem cells associated with accelerated ageing. Nat Cell Biol 10:452–459

Soldner F, Laganière J, Cheng AW, Hockemeyer D, Gao Q, Alagappan R, Khurana V, Golbe LI, Myers RH, Lindquist S, Zhang L, Guschin D, Fong LK, Vu BJ, Meng X, Urnov FD, Rebar EJ, Gregory PD, Zhang HS, Jaenisch R (2011) Generation of isogenic pluripotent stem cells differing exclusively at two early onset Parkinson point mutations. Cell 146:318–331

Suzuki K, Mitsui K, Aizawa E, Hasegawa K, Kawase E, Yamagishi T, Shimizu Y, Suemori H, Nakatsuji N, Mitani K (2008) Highly efficient transient gene expression and gene targeting in primate embryonic stem cells with helper-dependent adenoviral vectors. Proc Natl Acad Sci USA 105:13781–13786

Takahashi K, Yamanaka S (2006) Induction of pluripotent stem cells from mouse embryonic and adult fibroblast cultures by defined factors. Cell 126:663–676

Takahashi K, Okita K, Nakagawa M, Yamanaka S (2007) Induction of pluripotent stem cells from adult human fibroblasts by defined factors. Cell 131:861–872

Tiscornia G, Vivas EL, Belmonte JC (2011) Diseases in a dish: modeling human genetic disorders using induced pluripotent cells. Nat Med 17:1570–1576

Urbach A, Bar-Nur O, Daley GQ, Benvenisty N (2010) Differential modeling of fragile X syndrome by human embryonic stem cells and induced pluripotent stem cells. Cell Stem Cell 6:407–411

Vig J, Campisi J (2008) Puzzles, promises and a cure for ageing. Nature 454:1065–1075

Yang Y, Seed B (2003) Site-specific gene targeting in mouse embryonic stem cells with intact bacterial artificial chromosomes. Nature Biotechnol 21:447–451

Yang SH, Meta M, Qiao X, Frost D, Bauch J, Coffinier C, Majumdar S, Bergo MO, Young SG, Fong LG (2006) A farnesyltransferase inhibitor improves disease phenotypes in mice with a Hutchinson-Gilford progeria syndrome mutation. J Clin Invest 116:2115–2121

Yu J, Vodyanik MA, Smuga-Otto K, Antosiewicz-Bourget J, Frane JL, Tian S, Nie J, Jonsdottir GA, Ruotti V, Stewart R, Slukvin II, Thomson JA (2007) Induced pluripotent stem cell lines derived from human somatic cells. Science 318:1917–1920

Index

F.H. Gage and Y. Christen (eds.), *Programmed Cells from Basic Neuroscience*
to Therapy, Research and Perspectives in Neurosciences 20,
DOI 10.1007/978-3-642-36648-2, © Springer-Verlag Berlin Heidelberg 2013